WITHDRAWN

Manufacturing Planning

INDUSTRIAL ENGINEERING

A Series of Reference Books and Textbooks

Editor
WILBUR MEIER, JR.
Dean, College of Engineering
The Pennsylvania State University
University Park, Pennsylvania

Volume 1: Optimization Algorithms for Networks and Graphs,
Edward Minieka

Volume 2: Operations Research Support Methodology,
edited by Albert G. Holzman

Volume 3: MOST Work Measurement Systems,
Kjell B. Zandin

Volume 4: Optimization of Systems Reliability,
Frank A. Tillman, Ching-Lai Hwang, and Way Kuo

Volume 5: Managing Work-In-Process Inventory,
Kenneth Kivenko

Volume 6: Mathematical Programming for Operations Researchers and Computer Scientists,
edited by Albert G. Holzman

Volume 7: Practical Quality Management in the Chemical Process Industry,
Morton E. Bader

Volume 8: Quality Assurance in Research and Development,
George W. Roberts

Volume 9: Computer-Aided Facilities Planning,
H. Lee Hales

Volume 10: Quality Control, Reliability, and Engineering Design,
Balbir S. Dhillon

Volume 11: Engineering Maintenance Management,
Benjamin W. Niebel

Volume 12: Manufacturing Planning: Key to Improving Industrial Productivity,
Kelvin F. Cross

Additional Volumes in Preparation

Manufacturing Planning
Key to Improving Industrial Productivity

Kelvin F. Cross
Department of Industrial Engineering
Wang Laboratories, Inc.
Lowell, Massachusetts

MARCEL DEKKER, INC.　　New York and Basel

Library of Congress Cataloging-in-Publication Data

Cross, Kelvin F.
Manufacturing planning.

(Industrial engineering; 12)
Includes bibliographies and index.
1. Production planning. I. Title. II. Series.
TS176.C76 1986   658.5'03   85-29360
ISBN 0-8247-7324-1

COPYRIGHT © 1986 MARCEL DEKKER, INC. ALL RIGHTS RESERVED

Neither this book nor any part may be reproduced or transmitted in any form or by any means, electronic or mechanical, including photocopying, microfilming, and recording, or by any information storage and retrieval system, without permission in writing from the publisher.

MARCEL DEKKER, INC.
270 Madison Avenue, New York, New York 10016

Current printing (last digit):
10 9 8 7 6 5 4 3 2 1

PRINTED IN THE UNITED STATES OF AMERICA

Preface

This book will be a welcome aid for anyone who is involved with developing manufacturing manpower, equipment, and facility requirements based on marketing forecasts, productivity changes, or revisions in manufacturing strategy. As a step-by-step guide or as a desk reference, the book will be helpful to all levels of manufacturing management, industrial/manufacturing engineers and managers, management consultants, and students of manufacturing management or engineering.

    This work is unique in that it presents a comprehensive, organized approach toward rapidly and accurately developing medium- to long-range manufacturing requirements. Other books have not dealt adequately with this problem. Facilities planning books, like plant layout books, tend to emphasize optimal growth plans and layouts with the assumption that one has already determined the equipment and floor space requirements. Production planning books emphasize short-term production scheduling and inventory control, with secondary emphasis on short-term capacity planning. This text fills the gap between the two disciplines by providing a readily usable approach toward developing medium- to long-range manufacturing requirements.

Although manufacturing planning has always been a necessary part of doing business, its importance has increased. The emphasis on manufacturing planning in today's market can be attributed to increased product and process complexity; the rapid pace of technological change; increased competitive and economic pressures; and increased manufacturing strategy options. A systematic, organized approach to manufacturing planning will greatly enhance its effectiveness. By using the accurate, comprehensive manufacturing planning system outlined in this book, any company will improve its productivity and profitability. The use of proper planning techniques will achieve these goals by ensuring that

1. Manufacturing resource requirements are synchronized with production demands.
2. Management, marketing, and business strategies are based on a realistic assessment of the factories projected and their current capabilities.

While these factors alone make the effort devoted to manufacturing planning extremely cost-effective, this effort may easily lead to productivity gains in the short term by uncovering previously hidden areas of inefficiency.

This work details a step-by-step procedure for developing a manufacturing plan and an ongoing manufacturing planning function. Specifically, the techniques for defining, collecting, and manipulating the large volume of data are clearly organized and presented. In addition, there is a chapter on the application of computers to manufacturing planning, from the in-house mainframe to the off-the-shelf microcomputer. Analysis of proposed manufacturing requirements and alternative manufacturing strategies are discussed. Another chapter details how this manufacturing planning procedure can be used as a tool for analyzing current productivity. The procedure is intended primarily for the standard cost production environment, although many of the definitions and methods also apply to job shops and process industries.

Manufacturing planning done in some form is unavoidable, yet it is essential to the productivity of any factory. Done properly, with an understanding of its importance, the contribution to plant productivity will greatly exceed its costs. Used as a tool for strategic planning decisions, the impact on a company's profitability will be even greater. This potential for

Preface

manufacturing planning and the techniques used to achieve it are presented throughout the text in a thorough yet understandable and usable format.

Kelvin F. Cross

Contents

Preface   iii

1. INTRODUCTION TO MANUFACTURING ASSETS PLANNING   1

    1.1 Definition of Manufacturing Assets Planning   1
    1.2 Productivity and Profitability Planning   3
    1.3 History of Manufacturing Planning   9
    1.4 The Increased Need for Manufacturing Planning   10
        References   17

2. MANUFACTURING STRATEGY   19

    2.1 Introduction   19
    2.2 Corporate Philosophy and Strategy   20
    2.3 Corporate Strategy and Its Impact on Manufacturing   22

2.4	Integrating Manufacturing Strategy and Planning	29
2.5	The Current Operation	32
2.6	The Role of Manufacturing Capacity and Assets Planning (Systems)	32
	References	33

3. MANUFACTURING PLANNING, PREPARATION, AND ORGANIZATION   35

3.1	Introduction	35
3.2	Objectives and Goals	35
3.3	Centralized Responsibility	45
3.4	Manufacturing Planning as an Ongoing Function	47
3.5	Organization	52
	References	57

4. DATA COLLECTION   59

4.1	Importance of Data	59
4.2	Difficulty of Collecting Data	60
4.3	Variable Versus Fixed Data	65
4.4	Summary	66

5. PRODUCT DATA   67

5.1	Introduction	67
5.2	The Company's Products	68
5.3	The Products of the Factory or Manufacturing Department	73
5.4	Converting the Company's Product Forecast to a Factory Production Schedule	75
5.5	Product Design Changes and New Technologies	87
	Reference	89

6. MANUFACTURING PROCESS DATA   91

6.1	Introduction	91
6.2	Manufacturing Operations and Sequence	91

Contents   ix

    6.3 Labor Planning Data   99
    6.4 Equipment Planning Data   104

7. FACILITY DATA   109

    7.1 Introduction   109
    7.2 Equipment Square Footage   110
    7.3 Work-in-Process   114
    7.4 Indirect Areas   114
    7.5 Office Space   118

8. PRODUCTIVITY FACTORS   123

    8.1 Introduction   123
    8.2 Definition and Impact on Manufacturing
         Planning   124
    8.3 Collecting the Data   133
    8.4 Projecting Improvements   138

9. THE PLANNING PROCESS   141

    9.1 Introduction   141
    9.2 Requirements Planning Versus Tactical
         Planning   142
    9.3 Calculating Requirements   143
    9.4 Determining Strategy   145
    9.5 Rough-Cut Manufacturing Planning   146

10. GENERATING A PRODUCTION SCHEDULE   151

    10.1 Determining a Time Frame   151
    10.2 Converting the Product Forecast into a
          Factory-Built Plan   153

11. GENERATING MANUFACTURING REQUIREMENTS   165

    11.1 Capacity Requirements by Operation   165
    11.2 Work Center Requirements   178
    11.3 Total Production Line Requirements   180

11.4	Additional Requirements for Employees and Space	190

12. WAREHOUSE FACILITY PLANNING — 193

12.1	Introduction	193
12.2	Data Collection	194
12.3	Warehousing Requirements Planning	210

13. MANUFACTURING TACTICAL PLANNING AND IMPLEMENTATION — 217

13.1	Introduction	217
13.2	Evaluating the Requirements Plan	217
13.3	Revising the Requirements Plan	219
13.4	Presentation to Management	223
13.5	Implementation Planning	227

14. COMPUTER-AIDED MANUFACTURING ASSETS PLANNING — 235

14.1	Feasibility Analysis	235
14.2	Requirements Analysis	238
14.3	Utilizing the "In-House" Mainframe Computer	242
14.4	Microcomputer Applications	244

15. PRODUCTIVITY PLANNING: UTILIZING THE MANUFACTURING PLANNING SYSTEM — 261

15.1	Introduction	261
15.2	Auditing Productivity	262
15.3	Analyzing Productivity Improvement Potential	265
15.4	Planning Productivity	266

Index — 281

Manufacturing Planning

1
Introduction to Manufacturing Assets Planning

1.1 DEFINITION OF MANUFACTURING ASSETS PLANNING

Manufacturing planning is the process of determining future manpower, equipment, and facilities requirements. Primarily these manufacturing requirements must be based on marketing forecasts of future product demand. Other variables that impact future manufacturing requirements include changes in production method, product structure, and manufacturing strategy.

A medium- to long-range manufacturing plan defines when, where, and how much the requirements will change for manpower, equipment, and facilities. Profitability and productivity are then maximized and costs minimized by having the right assets available at the right time and at the right place. The same analysis and planning methods may also be used as a systematic, organized approach to assess present conditions, discover hidden costs, and improve short-term productivity.

The emphasis of this text, then, is on medium- to long-range manufacturing planning. Typically, medium to long range

Introduction to Manufacturing Assets Planning

means 6 months to 5 years, although some industries may have a need to project their requirements even further.

1.1.1 Versus Production Planning

This emphasis on the medium to long range is in stark contrast to the short-range orientation of production planning. A recent text on production control states, "The process of developing . . . (the) . . . kind of relationship between market demands and production capabilities day in and day out is the function of production planning, scheduling, and inventory control" [1].
   Manufacturing planning, in the context of this text, is quite different. It is not at all oriented towards day-to-day scheduling and controls. Rather than being concerned with short-term inventories and work in process, manufacturing planning is oriented toward future capacity and resource requirements. Also, the thrust of production control is on short-term material requirements, while manufacturing planning is oriented towards long-term asset requirements.

1.1.2 Versus Facilities Planning

Manufacturing planning should not be confused with facilities planning. Although they are very similar in definition, there are some key differences. Manufacturing planning involves quantifying future manpower, equipment, and facilities requirements. Facilities planning ". . . involves devising and determining how your plant should be physically orgainized to perform or produce" [2]. In other words, manufacturing planning emphasizes determining resource/asset requirements, while facilities planning emphasizes organizing those requirements.

1.1.3 Versus Capacity Planning

Capacity planning is another term that is used frequently. Although it also addresses manpower, equipment, and facilities, it emphasizes their short-term optimum utilization. In this sense, it is a function of production scheduling and control. Capacity planning involves accommodating variable production requirements with fixed capacity resources. Manufacturing planning, on the

Productivity and Profitability Planning

other hand, views capacity resources as variable to meet future production demands.

However, manufacturing planning methods can be used as a tool for capacity planning. Manufacturing planning methods can be employed to develop a comparison of projected asset requirements versus the assets on hand. Such a comparison will illustrate specific capacity constraints and bottlenecks that must be addressed.

1.2 PRODUCTIVITY AND PROFITABILITY PLANNING

Effective manufacturing planning will ensure that the right resources/assets are available at the right time and at the right place. Future productivity and profit potential will then be optimized due to organized factory growth and maximum manufacturing efficiency.

1.2.1 Cost Avoidance

Manufacturing planning is done to minimize future costs. It is an action that can be taken now to manage the future. An action now should minimize the need for a future reaction to past mistakes. Manpower, equipment, or facilities that are obtained before they are needed will create an unnecessary drain on working capital. In extreme cases, the acquisition of unneeded assets can cause the financial collapse of a corporation. For instance, a new facility may be purchased long before it is needed. This could effectively devastate a company's cash flow by tying up millions of dollars with little or no return on that investment.

Less dramatic but much more common is when production equipment is purchased but not used. For example, assume that the cost of a new piece of production equipment is $150,000 and that it costs an additional $50,000 to deliver and install the equipment. Also assume that this equipment was operational 6 months before it was required for production. This new equipment will essentially tie up $200,000 with no return for the 6 months it is unused. Depending on the size of the company, tying up $200,000 could have a major impact on its financial condition. Some companies simply cannot afford to take $200,000 out of its cash flow without a speedy return.

Another factor to consider is the interest on $200,000. Had it been left in the bank for 6 months at 10% interest, $10,000 would have accrued. By tying up $200,000 in unneeded equipment, this $10,000 of potential interest is lost. This loss would be even greater if there were more machines purchased, square footage utilized, and more people hired. However, the planner must be aware that there are cases where installing equipment 6 months early makes sense. Sometimes there are considerations that are not readily apparent, such as situations involving business strategy or tax considerations. Sound financial planning could be improved by a more accurate definition of when new equipment should be purchased, delivered, and installed, people hired, and new facilities acquired.

An even more costly error for a business to make is to acquire assets after they are needed. Time is always lost forever. If shipments are missed, they often cannot be made up. In a competitive market, the consumer will go elsewhere. In industries with a short product life cycle, such as computers, missing shipments on a new product may mean losing the market altogether. The potential loss is difficult to measure, but could easily amount to millions.

A more common scenario is one in which a piece of equipment with a long lead time is ordered too late; therefore, it is in production too late. For example, assume a 10% increase in production was held up for 13 weeks until the new equipment was on line. If the current line produces a profit of $200,000 per week, the 10% loss in profit would amount to $260,000. This example may be an oversimplification, but it magnifies the importance of having the resources to produce the product at the right time.

1.2.2 Operating Strategies and Decision Making

Manufacturing planning improves profitability by enhancing corporate decision making. It does this by ensuring that management, marketing, and business strategies are based on a realistic assessment of the factory's current and projected capabilities.

Manufacturing resources must be available at the right time, but it is also important to have them in the proper place. The productivity of any factory can be enhanced by the efficient placement of equipment, facilities, and manpower. In many factories, the layout of equipment and the flow of materials have

Productivity and Profitability Planning

evolved with little regard for long-range impacts. For example, new equipment was placed where it would fit.

Long-range manufacturing planning can provide a master plan for factory development. Facility additions and equipment placement decisions can be made in the context of an overall plan. Planned factory growth will help to ensure a more productive factory in the future and eliminate the need for some future cost reduction projects.

Manufacturing Management

A comprehensive manufacturing plan can be used as a guideline for manufacturing management. Short-term equipment, installation, and relocation decisions can be made in the context of the overall plan. A long-range plan should reduce the time needed to make short-range decisions.

Facilities management and plant engineering will also be aided by manufacturing planning. A medium- to long-range projection of equipment requirements can provide the data needed to effectively manage utility and service lines. For example, assume a manufacturer is installing a new piece of equipment that requires high pressure air, steam lines, 220 V electricity, and a drain. Also assume that by looking at the manufacturing plan, one could tell that within 3 years, three more pieces of equipment would be purchased. Knowing this, a plant engineer could require that the new plumbing lines be capable of handling an additional three pieces of equipment. This would greatly reduce future plumbing and installation costs.

Personnel Management

Personnel management strategies would be enhanced by manufacturing planning. A detailed manufacturing plan may illustrate the need for future manufacturing managers and professionals. Knowing this in advance, a properly oriented human resource development program can be implemented.

Marketing Management

A documented manufacturing plan will enable marketing to do its job better. Through understanding the capabilities and constraints of manufacturing, both current and projected, the marketing effort can be optimized.

Specifically, marketing provides a forecast of product demand. A manufacturing plan is then developed to ensure that the product is produced and that demand is met. If marketing does not meet its forecast, the factory's excess capacity will incur costs. On the other hand, if marketing can outsell what was forecasted, sales will be lost because of the inability to produce. The ideal situation is an equal balance between manufacturing and marketing.

A manufacturing plan will predict the impact of the marketing forecast. Through an understanding of the difference between manufacturing capacity and marketing demand, the business becomes manageable. Marketing's understanding of this balance may help reinforce the necessity of accurately forecasting sales and meeting those forecasts. Hence, manufacturing considerations may play a bigger role in forecast revisions.

Financial Management

Sound financial planning can also be enhanced through realistic assessment of the factory's current and projected capabilities. A manufacturing assets plan will enable the development of accurate and reliable budgets thereby assuring a stable business and maximizing profits.

In conjunction with financial planning, sound business decisions require a solid understanding of the manufacturing function. Only through an understanding of the current and projected manufacturing operation can alternative operating scenarios be evaluated. In particular, these studies may range from reevaluating make or buy decisions to studying the feasibility of offshore manufacturing.

Corporate Communications

Simply the act of developing a manufacturing plan will make a contribution to the organization's overall productivity. Developing a plan can help foster line-staff communications. Manufacturing planning will also generate communications from the top down, and back, involving many departments within the organization and reinforcing the attitude that we are "all in this together."

The documented manufacturing plan is also a concrete and specific goal. Done properly, it will be realistic and achievable,

Productivity and Profitability Planning

and will serve to focus action and motivate personnel. A manufacturing plan will also provide a yardstick for measuring achievement and progress.

A documented manufacturing plan can help minimize fear of the future and resistance to change. This is especially true if there is participation in the planning process.

1.2.3 Productivity Improvement

Short-Term Productivity

Although manufacturing planning is oriented toward the future, it may have an immediate and positive impact on the present. A comprehensive manufacturing planning effort will uncover previously hidden areas of inefficiency. Inefficiencies are exposed because every aspect of the interrelationship between the product and the process comes under scrutiny. Documenting this interrelationship will generate a keen understanding of its current makeup. Through this understanding, new ideas for improvement will also be generated.

There may be instances when a manufacturing plan indicates that a future production increase can be done with fewer people or machines than currently exist. If this is true, either the manufacturing plan is incorrect or the factory is underutilized. With an accurate plan, a detailed comparison of future projected resource requirements to existing resources will pinpoint areas requiring investigation. Operations that are overstaffed for current production requirements can be reduced. Equipment downtime, utilization, and other efficiency problems can be discovered and corrected. Inefficient use of valuable factory floor space can also be highlighted and corrective action taken where possible.

The manufacturing planning procedure can also be used as a productivity study. It is just as easy to generate a manufacturing plan with today's production rate as it is with a forecasted production rate. By using current rates instead of a forecast, the difference between actual and calculated manufacturing requirements will be clearly visible. This variance can then be used to define and prioritize an effective series of cost reduction projects.

8   Introduction to Manufacturing Assets Planning

Long-Term Productivity

Manufacturing planning may also help identify future productivity possibilities. For example, the current manufacturing method may be sufficient to produce relatively low production volumes. With a much higher production rate, a manufacturing plan may show that an intolerable level of work-in-process would occur. This would demonstate the need for an improvement in process flow, perhaps with material handling equipment or robotics. Once these needs are identified, steps can be taken to meet these needs at the most opportune time.

1.2.4 Planning Costs

The cost of developing a manufacturing plan or planning system is minimal in contrast to its return. Unfortunately, much of its return is intangible and cannot be measured accurately. It involves speculation to estimate the costs that would have occurred had planning not been done. It also involves speculation to predetermine the cost reduction ideas that might occur during the planning process.
    Developing a manufacturing plan is somewhat analogous to doing preventive maintenance: incurring costs now may avoid a major problem later. The costs incurred in developing a manufacturing plan, or planning system, are essentially a function of time. The major commitment of time is required by the planner or coordinator. In addition, time will also be required of others within the organization, such as corporate management, manufacturing management, production and product engineering, financial management, production control, and marketing personnel. Some businesses may find it necessary to retain the services of an outside consultant.
    A recent example will serve to put these costs in perspective. A comprehensive 5-year manufacturing plan was developed for a major consumer goods manufacturer. This $100 million company employs approximately 1500 people in its 500,000 square foot facility. Development of a detailed plan took approximately 6 months. This required the fill-time effort of an industrial engineer and about half of that time for an outside consultant. The total cost for developing this manufacturing plan could be ballparked at $60,000. $15,000 for a half year of industrial engineering effort, $30,000 for the consultants, and $15,000 for

the time required of others throughout the company. Most of this effort was spent collecting and organizing the base data. Once the data are collected and organized, regenerating a new plan is far less time consuming.

Even at $60,000 the effort more than paid for itself. In fact, costs were avoided by insuring the optimum timing and placement of new manufacturing resources. Developing this plan also resulted in some short-term benefits. During this study, $90,000 of unnecessary material handling labor was uncovered. Other areas of inefficiency were also uncovered and improvements recommended.

1.3 HISTORY OF MANUFACTURING PLANNING

Theoretically, manufacturing planning preceded manufacturing itself. Before establishing the first manufacturing plant, decisions had to be made on how, where, and by whom the manufacturing was to be done. Planning was probably intuitively done for the short range and not formalized—an art, not a science. It was not until the turn of the century that long-range planning came into vogue as a business technique. Most of these plans were still based on the instincts of the head of the business. These plans were conceptual and strategic in nature and rarely involved details.

It was not until the 1920s that long-range planning for manufacturing was even discussed in the literature. This was the first time that recognition was given to the problems of sales forecasting and long-range requirements planning. There was increased awareness of the distinct separation between sales and manufacturing. It became more widely recognized that manufacturing was responsible to the sales department. If the sales department could not sell a requested product, the sales department assumed the blame. While production forecasts were the responsibility of the sales department, producing those forecasts was the responsibility of manufacturing.

In the early 1930s long-range planning was still a minor concern of most companies. In a survey of 31 large organizations, only 2 had done any planning for their operations as far out as 5 years. Half of the companies planned their operations 1 year in advance [3].

In 1967-68—35 years later—a study was made of facilities planning in 302 well-established American manufacturing firms.

This study was referred to by Hales and Muther in their book *Systematic Planning of Industrial Facilities, Volume 1*. They said that it "indicated that over 60% followed this typical site planning approach: 'We plan each expansion as needed, but we always look ahead with a few years projection of specific needs and try to make the present facility plan fit into the next expansion or rearrangement.' Sixteen percent reported that they did not even do this, and only 22% indicated that they had made any overall site plan which they attempted to follow" [2].

Another survey of corporate planning practices was done by *Business Management* magazine in 1967. It indicated that half of the 101 companies surveyed were involved in long-range planning. Although manufacturing planning was not mentioned specifically, 67 of the companies answered yes to the question of whether or not operating managers make estimates of the manpower they need to reach company goals [4].

During the 1970s, long-range planning took on a significant role in American business. *Long Range Planning* magazine—the very existence of which is significant—summarized two studies of industry practice, one done in 1974 and the other in 1979 [5]. In both cases, 86% of the companies responding developed some type of written long-range plan (3 years ahead or more). In 1974 and 1979, 82% of the companies' plans included formal plans and budgets for plant expansion. Long-range planning of equipment acquisitions increased from 57% in 1974 to 68% in 1979. Not surprisingly, the use of computers and models in long-range planning increased from 47 to 61% in the same 5-year period [5].

This emphasis on long-range planning, particularly manufacturing planning, will probably increase throughout the decade. It is primarily due to the rapid pace of change facing our country and the world. Not only are advances in technology being made at an increasing pace, but also changes in social structure, world politics, the economy, regulations, and competitive pressures. As this pace quickens, the need for planning becomes more pronounced.

1.4 THE INCREASED NEED FOR MANUFACTURING PLANNING

Many changes are affecting the way business is conducted. Successful businesses adapt to these changes and incorporate them

The Increased Need for Manufacturing Planning

into their operating strategies. Adapting to change requires planning, and many companies are actively involved in the process of corporate strategic planning. Unfortunately, strategic planning is frequently perceived to be primarily a marketing or financial process. During the last few years, there has been an increasing emphasis on the role of manufacturing in the corporate strategic-planning process. Manufacturing is being recognized as a corporate strategic weapon rather than as a distinctly independent but necessary function. The company's strategic plans depend on manufacturing's ability to carry them out. Therefore, a successful strategic plan depends on a compatible manufacturing plan.

The importance of synchronized planning is amplified when a company introduces a new product. A business strategy might suggest that the marketing conditions would be right and the financial benefits great if a new product were to be introduced. Unfortunately, if that product could not be manufactured on time, the favorable marketing and financial conditions would be lost forever. Had the strategic planning process taken into account manufacturing capacity and capability, an alternate strategy could have been devised. In some cases, the alternative might include abandoning or never developing the new product in the first place. Along with timing, considerations of cost and quality could have a similar impact on corporate strategy.

A specific example of this can be seen in a medium-sized corporation's ill-fated entry into the office equipment manufacturing business. Nameless Corporation is a major manufacturer and distributor of office equipment supplies. With its distribution network and its additional experience selling Japanese office equipment, Nameless felt that they could successfully produce and sell their own office equipment. Marketing plans suggested that the timing would be right and that the rewards would be great. Unfortunately, the marketing strategy and manufacturing capability did not match. The result was a disaster. Manufacturing was unable to produce a quality product, at the right cost, within the time frame provided. Eventually, the program was canceled, although much too late. Nameless lost millions of dollars on product development and on acquired manufacturing capacity. An accurate, comprehensive, and detailed manufacturing plan would have avoided these problems or altered the strategic plan.

The success of a corporate strategic plan depends on the capacity and potential capability of the manufacturing operation. In many industries, determining these capabilities is a difficult process. In recent years there have been a number of factors that contribute to this difficulty.

1.4.1 Technology

Technological changes impact manufacturing by affecting both the product and the process. As these new technologies evolve, it has become increasingly necessary to review their impact on the manufacturing operation. What has changed recently is not that there are changes to both the product and the process, but rather the rate of that change. The future success of the manufacturing operation depends on incorporating these changes. Preparing for technological change requires planning.

Manufacturing planning, in particular, is affected by this onslaught of technology. In most industries, a major impact on manufacturing will come from obvious technological trends. Sophisticated technological forecasting probably will not be necessary. While these technological trends may be obvious, their impact on the manufacturing operation may not be. Developing an accurate picture of future manufacturing capability requires a detailed manufacturing plan.

Many products will be greatly altered by advances in technology. In some industries, existing products will be rendered obsolete and replaced with new, more advanced models. More commonly, companies will produce the same products for years. However, even in these companies the manufacturing process may be significantly altered due to new materials and operations used in producing these products. From a marketing perspective the product may remain the same, while from a manufacturing standpoint the product has completely changed. Examples of this difference in perspective can be seen in products ranging from sewing machines and steam irons to automotive dashboards. Today, most of these products are heavily dependent on the manufacture of plastic. Whether done in-house or by an outside vendor, somewhere metal-working operations were eliminated and replaced by the manufacture of plastic.

Examples can be cited in other areas. Many products, therefore manufacturing operations, are greatly impacted by such new technologies as advances in microelectronics, man-made fibers, genetic engineering, and ceramics, as well as metals and plastics.

The Increased Need for Manufacturing Planning 13

Advances in product technology are overwhelming. Even assuming that the product stays the same, the method of manufacture may be radically different in the future. Advances in manufacturing technology may reduce labor requirements, but increase dependence on supersophisticated automated assembly and fabrication equipment. A detailed manufacturing plan will enable these new processing technologies to be successfully managed and implemented. Some process technologies that have increased the need for manufacturing planning include robotics, automatic storage and retrieval systems, process controllers, and a variety of computer-controlled automated assembly, fabrication, and inspection equipment.

1.4.2 Shorter Product Life Cycles

Shorter product life cycles also result from the rapid pace of technological change. Some products—such as many complex high-technology items—may become obsolete within a year of their introduction; therefore, they must be delivered on time and be manufactured cost effectively. As a result, manufacturing planning becomes both necessary and more difficult. Correctly forecasting the resources required to produce these products is essential to the success of both the product and the business.

1.4.3 Product Complexity and Specialization

Product complexity and specialization also contribute to the difficulty of the manufacturing management. Although many products are complex due to their materials and construction, real complexity occurs when each product is tailored for each customer. This trend to customization can be seen in many industries. In the automotive industry, a customer can order a wide variety of options. The manufacturer will then build a car to that customer's specifications. Manufacturers in many diverse industries modify their standard manufacturing procedures to satisfy customer requirements.

Most businesses have become marketing oriented rather than manufacturing oriented. A manufacturing-oriented company has the attitude: "This is what we make; if you want it, buy it." A marketing-oriented company has the attitude: "If you want to buy it, then that's what we will make." Although oversimplified, one can see how today's marketing orientation complicates the manufacturing picture.

Alvin Toffler, in his book *The Third Wave*, sees this marketing orientation continuing and becoming even more sophisticated. He states, "Vast changes in the technosphere and the infosphere have converged to change how we make goods. We are moving rapidly beyond traditional mass production to a sophisticated mix of mass and demassified products. The ultimate goal of this effort is now apparent: completely customized goods made with holistic continuous flow processes increasingly under the direct control of the consumer" [6].

1.4.4 Competitiveness and World Economy

Competitive pressures and the worldwide business climate greatly affect manufacturing strategy. In particular, the Japanese are recognized as fierce competitors, primarily due to their manufacturing effectiveness. This manufacturing skill has caused American industry to renew its emphasis on manufacturing strategy. Alternative manufacturing strategies and business strategies must be considered in light of this worldwide competition. A company that understands its present and future capabilities is in a position to respond to this challenge. A detailed manufacturing plan will provide a foundation for developing and evaluating strategies.

1.4.5 Overseas Manufacturing

Overseas manufacturing has become an increasingly attractive alternative. The overseas option is attractive to many industries with a high labor content in portions of their manufacturing operation. In these instances, it is cheaper to send components abroad for manual assembly and then have them shipped back. A recent example of this can be seen in the actions of a major baseball manufacturer. For over 50 years, this company manufactured all of its baseballs and softballs within one facility. This included winding the center, cutting the leather, sewing on the covers, stamping seals, and packaging. During the 1970s, with increasing labor costs, inflation, and a decrease in profit margin, it became more profitable to ship the cut leather

The Increased Need for Manufacturing Planning   15

along with the centers to Haiti, have the balls stitched there and then shipped back for stamping and packaging.

Besides cheap labor, overseas manufacturing is also considered with increasing frequency due to tax and financing advantages. For example, Puerto Rico has attracted many computer manufacturers due to its tax advantages. Such companies as Honeywell, Wang, and Digital all have manufacturing operations in Puerto Rico.

A glaring example of the impact of government incentives can be seen in the case of the DeLorean motor company. In effect, a bidding war developed in an effort to attract DeLorean's manufacturing operation. Puerto Rico lost this bidding war to Ireland when the British government offered a lucrative financing package, in addition to tax advantages.

As this trend toward a more global economy advances, a company that maintains an accurate perspective on its manufacturing capabilities as well as its limitations can develop a sound and competitive manufacturing strategy. Even without considering the global economy or product design changes, the manufacturing operation itself is impacted by a number of factors that increase the need for manufacturing planning.

1.4.6 Cost Increases

Equipment, facilities, and capital are extremely expensive, requiring careful analysis prior to their acquisition. Once purchased, both facilities and equipment become fixed assets. They are not readily convertible to money or easily resold. Expensive equipment immediately becomes used and depreciates. Facilities are physically fixed and cannot be changed easily. Whether purchased or leased, equipment and facilities are usually acquired for the long term. They also must show a financial return for the duration of that term. The need to meet manufacturing goals, combined with the expense involved, creates a need for sound study.

Although equipment and facility acquisitions always require sound study, current conditions make this need even more pronounced. Whether due to inflation, regulations, or other factors, the real cost of the acquisition of land and the construction of facilities has increased dramatically over the last decade. Production equipment has become both more sophisticated and more expensive. To compound this problem, interest rates have risen to new heights and will probably remain relatively high for the foreseeable future. Through an accurate manufacturing plan, acquisitions can be made in a cost-effective manner congruent with company strategy.

1.4.7 Equipment Lead Time

Sophisticated production equipment is not only expensive, but, in many cases, its acquisition takes time. Some production equipment requires as much as 1 year between the time it is ordered and the time it is installed and is operational. As production equipment becomes more sophisticated and specialized, this situation may become even more prevalent. Obviously, planning is required to have each piece of equipment in production at the right time. Planning must also take into account the time needed to train workers to run the equipment.

1.4.8 The Changing Work Force

The role of the worker in manufacturing continues to evolve in a way that necessitates a greater emphasis on manpower planning. Much of the manual labor is either being done overseas or is being automated. Manufacturing jobs will become more skilled by requiring the operation of sophisticated equipment and the use of computer controls. Therefore, production workers will require extensive training. Sophisticated equipment will require the hiring and training of these worker swell in advance of the time they are needed for full production. As manufacturing workers are hired less because of their physical dexterity and more because of their mental contribution, more time will be required during the hiring process. The time required to find the right person is in addition to the time required to train that worker once he or she is hired. Again, one way of dealing with this trend is to project long-term manpower requirements and to plan for their acquisition.

1.4.9 Overcrowded Facilities

Facilities that are approaching capacity require manufacturing planning. In rapid-growth, high-technology industries, capacity expansions must occur frequently. Even moderate-growth industries may find themselves forced to consider facilities expansion. Reluctant to expand in this turbulent economy, some companies have overstaffed their current facility beyond its efficient capacity. An overcrowded facility forces belated yet critical attention to be placed on manufacturing planning.

REFERENCES

1. P. Niland, *Production Planning, Scheduling, and Inventory Control*, MacMillan, Toronto, 1970.
2. L. Hales and R. Muther, *Systematic Planning of Industrial Facilities*, Vol. 1, Management and Industrial Research Publications, Kansas City, Kansa, 1979.
3. P. Holden, L. Fish, and H. Smith, *Top Management Organization and Control*, Stanford University Press, Stanford, Stanford, Calif., 1941.
4. Ed., How 101 companies handle corporate planning, *Business Management*, September 1967.
5. W. Boulton, S. Franklin, W. Lindsay, and L. Rue, *Long Range Planning*, 15:1 (1982).
6. A. Toffler, *The Third Wave*, Morrow, New York, 1980.

2
Manufacturing Strategy

2.1 INTRODUCTION

Understanding the importance of strategy is a must for effective planning. A well-conceived strategy provides the framework and direction for operations planning; a defined mission for the manufacturing operation. The operation and its related plans must support the corporation's overall business and operating strategies.

It is important to comprehend the relationship between corporate strategy and manufacturing. In particular, the impact of corporate strategy upon the design and method of manufacturing must be recognized. The corporate objectives for manufacturing can significantly impact process design and, therefore, the assets required.

The relationship between corporate strategy and manufacturing operations planning is illustrated in Fig. 2.1. An effective operation results from the integration of techniques, technologies, trends, and operating parameters in a way that best supports the overall strategy of the business. Combined with an

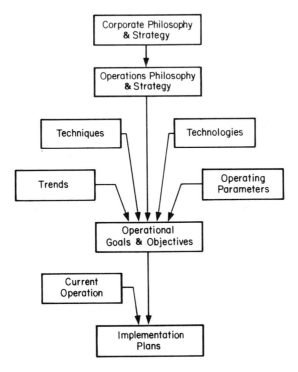

Fig. 2.1   Integrated operations planning.

understanding of current operations, the goal of an effective operation can be achieved through implementation plans.

2.2 CORPORATE PHILOSOPHY AND STRATEGY

In some companies the manufacturing function is perceived as a necessary evil. As an evil entity its strategic potential is ignored. Even more destructive is the tendency to misdirect the manufacturing function. For instance, in many cases the only strongly stated (and rewarded) objective within manufacturing is to minimize costs. This unfortunate attitude neglects the potential for manufacturing to make a major contribution to the overall objectives of the business. Understanding when other

Corporate Philosophy and Strategy

objectives for manufacturing are equal to or more important than cost is essential to sound planning.

The objectives for manufacturing must be understood within the context of the company's business. Each company, and their manufacturing function, exists to serve a market. How a company defines both its market and its approach for succeeding in that market is its corporate philosophy and strategy.

The content of a corporate philosophy was discussed by Ouchi in his book *Theory Z*. He states, "A corporate philosphy must include (1) the objective of the organization, (2) the operating procedures of the organization, and (3) the constraints placed on the organization by its social and economic environment. It specifies not only the ends but the means" [1].

On the other hand, in the book *Top Management Strategy*, Tregoe and Zimmerman discussed corporate strategy as a ". . . vision directed at what the organization should be, and not how the organization will get there." They defined strategy as ". . . the framework which guides those choices that determine the nature and direction of an organization. Those choices relate to the scope of an organization's products or services, markets, key capabilities, growth, return, and allocation of resources" [2].

The point is that it should be clear, to everyone within the organization, what the company is trying to achieve. Only through clear, communicated, and understood objectives can the organization proceed in a unified direction. Tregoe and Zimmerman described nine basic strategic areas [2]. At any given time in a corporation's life only one of these strategic areas should be pursued as the corporation's "driving force." Specific strategic policies and guidelines should then support the driving force of the company. The nine strategic areas (or driving forces) are listed below:

1. Products offered
2. Market needs
3. Technology
4. Production capability
5. Method of sale
6. Method of distribution
7. Natural resources
8. Size/growth
9. Return/profit

Although the application of these nine driving forces may appear obvious, this is not always the case. For instance, all high-technology companies are not necessarily technology driven. Until recently, a good example of a technology-driven company was Polaroid. Like all technology-driven companies, Polaroid has specialized in, and kept current in, a particular field of technology. In this case, the technology of instant photography was perfected and sold. Polaroid has offered state-of-the-art products, related to their technological expertise, to a variety of markets. Inexpensive cameras (but more importantly film and batteries) are sold to the consumer market. They also sell I.D. picture machines (and film) for commercial applications: licences, I.D. badges, etc. However, lately Polaroid has been changing its strategic direction. They have begun to market Japanese videotape. Marketing videotape, which is produced by others, does not take advantage of Polaroid's technological expertise. It does indicate Polaroid's desire to shift from being technology driven to either market or method of sale driven.

The driving force of a company will definately impact the manufacturing operation. A company driven by its production capability, such as U.S. Steel, would consider producing any product so long as it could be efficiently produced on its existing process. Therefore the manufacturing operation must be designed primarily for efficiency. The attitude is "This is what we make, if you want it—buy it."

In a company driven by market needs, the manufacturing operation must be flexible. The company will produce whatever products are needed to satisfy a particular market. The attitude toward their market is "If you want to buy it, we'll make it."

While these characterizations are generalized and oversimplified, they demonstrate the importance of strategy. Only through a top-down communication of the corporate strategy can the various organizational functions, such as marketing and manufacturing, be supportive.

2.3 CORPORATE STRATEGY AND ITS IMPACT ON MANUFACTURING

2.3.1 Corporate Manufacturing Strategy

The manufacturing function can make or break a company, depending on the extent of support it gives to the corporate

Corporate Strategy and Its Impact on Manufacturing

strategy. In successful companies, the corporate strategy is defined, communicated, and understood. Manufacturing, along with the other various organizational functions, then develops its own more specific operating strategy.

The corporate strategy implicitly defines the general performance criteria on which the manufacturing operation should be measured. The manufacturing operations strategy more explicitly specifies the performance criteria. In a sense, the manufacturing operations strategy is an interpretation of the corporate strategy.

Corporate strategy can influence manufacturing directly. For instance, assume that a part of a company's stated objective is "to provide customer-tailored, high-quality . . . products." The term "customer tailored" implies building the product to each customer's specifications. This simple declaration could eliminate from consideration any idea that would inhibit the flexibility of the manufacturing operation. Similarly, a company with a specific strategy to provide ". . .the best delivery time in the business" will accept the relatively high cost of this effort. Any new manufacturing method/system that is implemented solely to reduce costs, while compromising the ability to deliver, would likely fail or cause the business to fail. A successful manufacturing operation must support the overall corporate philosophy and strategy, as well as the bottom line.

In some cases, the manufacturing strategy may imply or specifically state operational guidelines. For instance, based on the corporate strategy and driving force, there may be guidelines regarding plant size or staffing levels. Either through experience or management style, an optimal factory size or manageable work-force level may become strategy. Any new plant must be designed to fall within the established strategic guidelines.

2.3.2 Product Manufacturing Strategy

Manufacturing strategy is also product dependent. There is an intermediary link between the overall corporate strategy and the more specific manufacturing strategies. That link can be found in the marketing strategies for each major product line. Therefore, a manufacturing strategy, which supports the marketing strategy, should be defined and implemented for each major product line and facility. A manufacturing strategy should spell out what the operation must be good at. No manufacturing operation can excel at everything; there must be a recognition of the important criteria. For instance, in the Polaroid example, the manufacturing strategy for producing inexpensive instant cameras would be quite different than the manufacturing strategy for producing commercial I.D. cameras. One is a high-volume, low-margin, low-complexity product. The other is low volume, highly complex, and perhaps a customer-configured product. In any case, the production operation for each product line must be designed to support the product's marketing strategy, as well as the overall strategy of the corporation.

There are some generic guidelines that may be helpful when trying to define a manufacturing strategy. These guidelines relate to the nature of the product that is to be produced and the appropriate strategy for its production. A product's characteristics and the stage of its life cycle can be primary determinants of the appropriate manufacturing strategy.

2.3.3 Vertical Versus Horizontal Integration

A company's total manufacturing operation is structured by its method of integration. A manufacturing operation can be integrated vertically, horizontally, or some mixture of the two. Vertical integration means that manufacturing departments are segmented by the product. Horizontal integration means that manufacturing is segmented by process. In other words, a product-focused plant is integrated vertically, while a process-focused operation is integrated horizontally.

In a horizontally integrated or process-oriented manufacturing environment, the operation is segmented by production method. Production management hierarchy is also delineated by manufacturing process. A process orientation is applicable to companies with complex yet distinct processing methodologies. Processing departments are usually distinguished by their capital equipment requirements and specialized technology.

A product-focused, or a vertically integrated manufacturing organization, is segmented by product. Vertical integration is usually applied when flexibility and innovation are important. Also, a product focus is particularly applicable when the production process is not capital intensive, technologically sophisticated, or overly complex.

The method of organizing the manufacturing function is further complicated by the option to combine both a product and

Corporate Strategy and Its Impact on Manufacturing 25

process orientation. For example, the first processing step for all of the company's products may involve a sophisticated technology or a specialized technique. For example, a rubber products company might have a centralized plant for extruding and mixing rubber compounds. In this case, the initial processing of the rubber compound is segregated and consolidated into one manufacturing department as a "feeder" plant or department. The first processing step will supply partially completed material to the specific product lines. In this scenario, the individual product lines would then independently complete the processing of their product. In the rubber company example, the extruded compounds might then be delivered to separate plants for further processing into toys, shoe soles, automobile bumpers, etc.

Whatever the approach, it is important to recognize that decisions regarding horizontal versus vertical integration are strategic. These decisions should not be based solely on considerations of manufacturing efficiency. The method of organizing the manufacturing operation should be primarily influenced by the corporate and product strategies. For instance, process-focused or vertically integrated manufacturing is usually more appropriate for technology- and natural-resource-driven companies. Conversely, a manufacturer driven by a "market needs" orientation would probably benefit from a horizontally integrated manufacturing organization.

Hayes and Schmenner argued that each manufacturing facility should be either product or process focused. Focus is important for manufacturing effectiveness. They stated, "Our position is not that both product and process focus cannot exist within the same company, but simply that separating them as much as possible will result in less confusion and less danger that different segments of manufacturing will be working at cross purposes" [3].

2.3.4 Product/Process Life Cycle

The method of integration, as well as the specific means of production, can be influenced by the product/process life cycle. When a product is initially introduced, it may be produced in small lots in a job shop. As the product matures, it may become more of a commodity and sold on the basis of price. At this stage, the manufacturing process is usually streamlined as a continuous-flow, high-volume operation.

This relationship, between product and process life cycles, is illustrated in Fig. 2.2. This matrix was portrayed by Hayes and Wheelwright in the *Harvard Business Review* to demonstrate the relationship and its strategic implications [4]. Specifically, a company's competitive posture is impacted by where it is positioned on this matrix relative to its competitors.

Regarding product decisions, Hayes and Wheelwright state, "A company may choose a product or marketing strategy that gives it a broader or narrower product line than its principal competitors. Such a choice positions it to left or right of its competitors, along the horizontal dimension of our matrix." Regarding process decisions, they went on to say, "Having made this (product) decision, the company has a further choice to make: Should it produce this product line with a manufacturing system . . . that will permit a relatively high degree of flexibility and a relatively low capital intensity? Or should it prefer a system that will permit lower cost production. . .? This choice will position the company above or below its competitors along the vertical dimension of our matrix."

2.3.5 Complexity Versus Volume Matrix

Understanding the relationship between product complexity and volume may also prove useful. In a company with a numerous variety of products, it may be helpful to categorize the products by their manufacturability. Product-to-process and product-to-plant alighment decisions can then be made with an adequate understanding of which products belong together.

Fig. 2.3 is complexity versus volume matrix for a mini-computer manufacturer. In this case, "complexity" has been equated with the standard hours needed to produce the product. "Volume" has been classified in terms of units per week. Note that the mini-computers are a low-volume, high-complexity product. However, the same company must produce high-volume, low-complexity terminals to accompany their mini-computer products.

This difference in product types may necessitate a difference in the means of production. However, complexity and volume are only two parameters by which products can be grouped. In other companies it may be more important to consider other parameters, such as product size, profit margin, material content, and uncertainty. A two-dimensional matrix also may be inadequate.

Corporate Strategy and Its Impact on Manufacturing 27

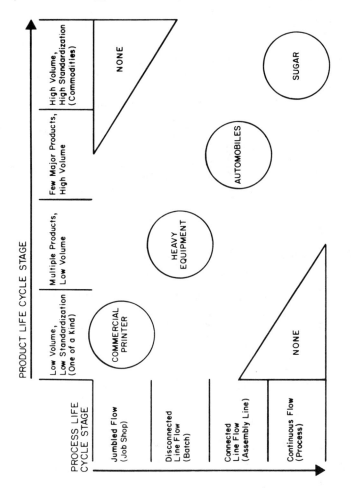

Fig. 2.2 The product/process matrix. (From Ref. 3.)

2.3.6 Complexity Versus Uncertainty Matrix

In a discussion of Japanese manufacturing strategies, Fitzpatrick and Puttock presented a complexity versus uncertainty matrix, displayed in Fig. 2.4 [5]. Each quadrant of their manufacturing strategy matrix describes one of four manufacturing situations. Within each quadrant the type of process, organization, and information method is described. Their main point in presenting the matrix was to explain Japan's success. Japan has done particularly well in high-complexity, low-uncertainty situations. They have focused on what they want to be good at, and they have succeeded.

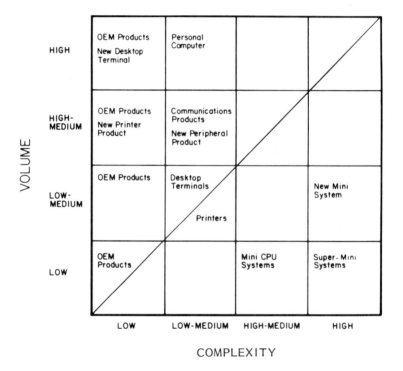

Fig. 2.3 Complexity versus volume matrix. A clustering of products means a similar manufacturability profile. The top of the diagonal line denotes products applicable to production line assembly, while the other products are most suited to a job shop process.

	Complexity	
Uncertainty	High	Low

High	Process:
	  Flexible
	Organization:
	  Strong
	  Centralized
	Information:
	  On-line
	  Coupled
	  Mechanized
		Process:
		  Adaptable,
		  Short lead time
		Organization:
		  Centralized
		  Responsive
		Information:
		  On-line
		  Decision oriented

Low	Process:
	  High variety
	  High productivity
	Organization:
	  Local control
	Information:
	  Uncoupled
	  Mechanized
		Process:
		  High productivity
		Organization:
		  Local control,
		  Protected
		Information:
		  Uncoupled
		  Optimized

Fig. 2.4 Complexity versus uncertainty matrix. (From Ref. 4.)

It is important that a company have an awareness of its manufacturing position in terms of the complexity/uncertainty matrix. It is especially important that each manufacturing facility be focused on only one quadrant. Although companies operate in more than one quadrant at a time, it is impossible to manufacture competitively in two quadrants using the same facilities due to the inevitable loss in performance and productivity.

2.4 INTEGRATING MANUFACTURING STRATEGY AND PLANNING

When planning future manufacturing operations there are numerous trade-offs and decisions to be made. Well-conceived corporate and manufacturing strategies provide the framework for making these decisions.

In the Integrated Operations Planning process (Fig. 2.1), these decisions usually involve choices among alternative operating techniques and specific manufacturing technologies. Primarily, the chosen operating methods must mesh with the company's operating strategy. However, the decisions will also be constrained by the operating parameters of the business and the realities of socioeconomic trends.

An effective operations plan results from the successful integration of these factors. Only through a portrayal of what the future operation should look like can the steps needed to achieve that goal be established. The factors, which must be integrated with strategy, are the trends, techniques, technologies, and operating parameters.

2.4.1 Trends

There are numerous social and economic trends that are relevant to manufacturing operations planning. Integrating these trends into the planning process, and therefore the future operation, could be critical to the success of the operation. For example, there are some major trends regarding the business of manufacturing and its work force. Business trends can strongly influence an operations plan. Any operations plan should consider the relative effectiveness of its competitors' future manufacturing operations. Effectiveness does not mean just cost effectiveness. There are trends, regarding manufacturing effectiveness, related to product quality, deliverability, and customization, as well as costs. Manufacturing is changing. Specifically, in response to product issues, there are business trends regarding automation and overseas manufacturing which cannot be ignored.

An integrated manufacturing operations plan should also consider the impact of a changing work force. Workers are becoming more highly educated and their expectations for rewarding work are greater than ever.

The combination of these trends creates a challenge. Manufacturing jobs must be designed to match the education and expectations of the work force; not because it is nice for the workers, but because good job design makes good business sense.

As manufacturing jobs are automated, and sent overseas, the nature of direct manufacturing jobs will change. Each worker will control more material and more capital equipment. For instance, one worker might control a 50,000-ft^2 factory full of robots. In essence, each worker will become more important to the success of the manufacturing operation. It is in the company's best interests to have motivated employees who are committed to company goals.

Integrating Manufacturing Strategy and Planning

2.4.2 Techniques

There are numerous alternative means of operating a manufacturing operation. The physical production operation as well as the control and management of that operation can be organized by a variety of techniques. There is no one right technique, yet some techniques are more appropriate than others for a given situation. In this case, "technique" refers to a mode of operation, as the examples below suggest:

Production methods alternatives: work cell, unit build, progressive assembly
Management style: incentive systems; quality circles
Control systems: MRP, Kanban; inspection procedures

2.4.3 Technologies

In the context of operations planning, numerous new technologies will provide a new means to produce the same product. The introduction of technological advances into the factory is essential and inevitable. To many, the integration of these technologies will create the "factory of the future." However, it is not that simple. The "factory of the future" will occur only through the integration of manufacturing and control techniques that are strategically innovative. A new technology must be employed as a means of achieving one's goals; it should not become a goal in and of itself.

However, all too often the "factory of the future" is perceived to be solely dependent upon the application of technological solutions. Frequently, "technology" is a solution looking for a problem, when it is the problems that must be understood. Primarily, an understanding of these problems, trends, and techniques is needed to put technological solutions in the proper perspective. It is this perspective and the realities of industry which must be incorporated into the operations planning process.

Through a strategic perspective it becomes possible to maximize the potential benefits of technology. Innovations in automation, robotics, and computerized control(s) systems are the tools of the factory of the future. Utilized in the proper context, these tools provide the means to an end, but are not an end in themselves.

2.4.4 Operating Parameters

In addition to the strategic objectives that must be achieved by the manufacturing function, the operation must also be designed with regard to other, more specific parameters. These parameters include capacity constraints, production forecasts, laws/ regulations, and corporate policies. These parameters are the hard realities of operations planning. Any plan for a new production line must at least address these issues.

Any operations plan depends upon utilizing operating assumptions, policies, and decisions. In some cases, these parameters are subject to change. Sometimes the requirements planning process itself may expose the need to alter a previously "given" operating parameter.

Manufacturing operations planning and design requires an integrated approach. Planning and design that is guided by the "big picture" is destined to be the most successful. Less time will be spent on the planning process itself. More importantly, the time that is spent on planning will be focused on achieving the major objectives of the corporation.

2.5 THE CURRENT OPERATION

A picture of the ideal future operation can come from integrated planning. However, achieving the ideal operation usually requires one more step. It is essential to develop realistic and achievable implementation plans.

Implementation planning must take into account the current operation. The only exception is when designing an entirely new operation from scatch. Any change in the current operating method can be disruptive. A disruption that will adversely affect production, even for a short time, is in many cases unacceptable. An implementation plan must strike a balance between achieving the ideal operation in the long term, and disrupting the operation in the short term.

2.6 THE ROLE OF MANUFACTURING CAPACITY AND ASSETS PLANNING (SYSTEMS)

The manufacturing capacity and assets planning approach, as described in this text, is a tool for understanding the operation.

As such, it provides information that is essential to the relevance and success of the integrated operations planning process.

An assets plan can define the operating parameters to be used in integrated planning. The product forecast, and the assets that are needed to manufacture the forecast, are operating parameters. While the manufacturing strategy qualitatively defines objectives, these parameters quantitatively define objectives.

The assets planning process is also especially useful for establishing a profile of the current operation. This profile, integrated with a profile of the ideal operation, establishes the groundwork for developing realistic implementation plans.

The assets planning process is also a feedback mechanism. An assets plan provides the opportunity to assess the impact of various production forecasts, changes in product mix, allocation of production to alternative plants, etc. Through this feedback, sound business decisions can be made and parameters established.

REFERENCES

1. W. Ouchi, *Theory Z*, Addison-Wesley, Reading, Mass., 1981.
2. B. Tregoe and J. Zimmerman, *Top Management Strategy*, Simon & Schuster, New York, 1980.
3. R. H. Hayes and R. W. Schmenner, How should you organize manufacturing?, *Harvard Business Review*, January/February 1978.
4. R. H. Hayes and S. C. Wheelwright, The dynamics of product-process life cycles, *Harvard Business Review*, March/April 1979.
5. Fitzpatrick and Puttock, Manufacturing strategies—lessons from Japan, *Production Engineering*, October 1983.

3
Manufacturing Planning, Preparation, and Organization

3.1 INTRODUCTION

Developing a comprehensive manufacturing plan requires preparation and an organized approach. Both must be done to ensure the usefulness and success of the manufacturing plan. A successful manufacturing plan will support the objectives of the business. It should be auditable, readily understood (therefore useful as a reference), and, obviously, accurate.

3.2 OBJECTIVES AND GOALS

Each company has its own objectives, goals, and critical areas of concern which the manufacturing plan must address. Therefore, it is imperative that those involved in generating the manufacturing plan have a sound understanding of the objectives. Decisions regarding the various manufacturing strategy alternatives that will crop up during the planning process can be made then with the proper perspective. Unfortunately, in many companies the objectives are not documented or clearly defined. Ideally, they should be documented, not just to support a manufacturing plan, but also as a guideline for management decisions.

Manufacturing planning must concern itself with a number of objectives being pursued by various functions throughout the organization. Not only do these functions perceive the organization's objectives differently, but they also have their own objectives for manufacturing planning. For example, the senior corporate executives will have a different set of concerns than manufacturing management. Likewise, the objectives and concerns of the business strategists will be quite different than those of a production engineering group.

First, it is important to understand everyone's perspective and operating objectives, especially with regard to manufacturing. Most likely this understanding will lead to the discovery of conflicting objectives about what the manufacturing plan should be designed to achieve. Such conflicts should be resolved before a manufacturing plan is developed. An effort must be made to ensure that future discussions of the manufacturing plan are not unnecessarily complicated or clouded by an underlying disagreement on objectives. It is better to resolve these conflicts now, rather than dispute the manufacturing plan later.

Second, it is important to understand who specifically is requesting the manufacturing plan and why. Depending on the requestor's objectives, certain aspects of manufacturing planning might receive increased attention while other aspects are ignored. Also, the plan can be summarized in such a fashion that the objectives of the user are met.

3.2.1 Corporate Management

One function of corporate management is to establish the overall operating philosophy of the business. Documentation of the operating philosophy and strategies will be a major benefit to the manufacturing planner. Such a document will provide a guideline for evaluating alternative manufacturing strategies. Consideration will then be given only to those alternatives that support the documented policy. This documentation of corporate philosophy and strategies should be developed by top management or the owners of the business. The content of a corporate philosophy is discussed by Ouchi in his book *Theory Z*. He states, "A corporate philosophy must include (1) the objectives

Objectives and Goals

of the organization, (2) the operating procedures of the organization, and (3) the constraints placed on the organization by its social and economic environment. It specifies not only the ends but also the means, [1]. Although this document may consist of only a few pages, the benefit to manufacturing planning and the organization could be substantial. For instance, assume that a company's stated objective is "to manufacture customer-tailored, high-quality products." The term "customer-tailored" implies building the product to each customer's specifications. This simple declaration could eliminate from consideration any idea that would inhibit the flexibility of the manufacturing operation.

Similarly, a company with a specific strategy of providing "the best quality level in the business" will accept the relatively high cost of this effort. Any new manufacturing method/system that is implemented solely to reduce material costs or production costs while compromising the "big picture" will fail or cause the business to fail. A successful manufacturing plan must support the overall corporate philosophy as well as the bottom line.

Corporate executive management is concerned with overall direction and general administration of the business. Corporate executive management is usually interested only in manufacturing planning when its outcome could have a significant impact on the direction of the business or on its financial status. It is not usually concerned with when or where equipment will be placed but, more to the point, how much will it cost and will the production capacity be there when it is needed.

A detailed manufacturing plan can provide executive management with a projection of future manufacturing capability and the costs which will be incurred obtaining that capability. Although the detailed manufacturing plan is essential to the accuracy of these projections, it is unlikely that upper management will want anything but a summary report. Executive management usually requires this information for one of two reasons: (1) it has made a decision on corporate direction and needs a manufacturing plan to support it, or (2) it is considering a new direction but, before deciding, needs to know its impact on manufacturing. In either case, the executives may not need much more than a page of information, yet to ensure that that page is accurate, a comprehensive manufacturing plan must be developed.

It is also essential to keep in mind that when executive management requests a manufacturing plan, it may really wind up wanting more than one. For example, the first manufacturing plan may show that manufacturing can meet the proposed objectives but it will cost too much. Corporate management will then ask that the plan be reconstructed on the basis of new operating assumptions and a new definition of the manufacturing objective. When a manufacturing plan is developed for corporate management, the regeneration of that plan is inevitable, especially when one considers their perspective of the manufacturing operation. Manufacturing is just one facet of the business. It must be manipulated or managed in a way to support the overall objectives of the business. To corporate management, manufacturing is a means to an end, not an end in itself. The end result is what is important. A manufacturing plan will enable these executives to make educated decisions regarding the manufacturing operation. When it is successfully integrated into business plans, a manufacturing plan will help to define the objectives of the business as well as achieve those objectives.

3.2.2 Business, Financial, and Marketing Management

The business, financial, and marketing organizations of a corporation usually must develop and achieve objectives that are similar to, and supportive of, the corporate philosophy. The major difference is that these organizations convert general corporate objectives into specific and achievable goals for their departments. For example, if the corporate philosophy required producing a product with an unexcelled level of quality, the marketing organization must implement a specific plan to meet this objective. Defining quality in terms of the market for each product and where that market is is the responsibility of the marketing department. Once departmental objectives are detailed and accepted by upper management, they should be communicated throughout the organization. No department is an island. Obviously, manufacturing management and manufacturing engineering would have to know of the quality requirements begin defined by marketing. Likewise, with finanacial objectives, their achievement requires the dedicated efforts of a number of departments.

A number of departments may have objectives that will impact manufacturing. Understanding and incorporating these objectives into the manufacturing plan is essential to the usefulness of the plan. Of particular importance is the relationship

Objectives and Goals

between marketing and manufacturing. The marketing department may have objectives that support corporate strategy yet also impact manufacturing. Shapiro, in an article in the *Harvard Business Review*, highlights eight areas where marketing and manufacturing objectives may conflict and where cooperation is necessary:

1. Capacity planning and long-range sales forecasts. If capacity is too low, management is upset because the company is losing sales; if capacity is too high relative to sales, management is upset because costs are too high. Since it takes time to significantly alter a plant's capacity, it is imperative that marketing make the effort to provide the most accurate forecast possible.
2. Production scheduling. The marketing function may be very concerned with the time it takes from when a customer gives an order until the time he receives the product. Objectives regarding this lead time should be defined and understood by both groups.
3. Delivery and physical distribution. Again, delivery time may be critical. Marketing needs may dictate inventory levels at various finished goods stocking locations.
4. Quality assurance. Manufacturing methods, material handling, and inspection operations may be determined in part by quality issues.
5. Breadth of product line. It is important to understand the marketing strategies regarding providing variations of each product.
6. Cost control. Obviously, marketing objectives for product cost will impact the manufacturing operation as well as the manufacturing plan.
7. New product introduction. While new end products should be a part of the marketing forecast, there may be marketing objectives regarding design changes on existing products. New product information could be a major component of the manufacturing plan.
8. Adjunct services such as spare parts inventory support, installation, and repair. Marketing may have goals regarding the production of spare components or the shipping of spare parts which manufacturing will have to produce.

Getting documentation of the corporate philosophy and the various departments' operating objectives that impact manufacturing may not be easy. Unfortunately, in many companies these operating objectives and the corporate philosophy are not well documented, but this does not mean that they do not exist. In many companies some of these objectives exist as part of the corporate climate or atmosphere. There are a lot of unwritten laws about how, where, and for whom the company will operate. Efforts should be made to document some of these unwritten laws. They should be used later and documented as part of the operating assumptions when the manufacturing plan is developed.

The main concern of marketing and other organization departments is that manufacturing support their goals. Although these departments are unlikely to directly request a manufacturing plan, they will have interest in such a plan. Marketing's primary concern should be that the plan show future manufacturing capacity to meet the marketing forecast. The primary concern of financial management should be that the marketing forecast be achieved in a cost-effective manner. A detailed manufacturing plan could provide an early warning signal regarding potential cost or capacity problems.

3.2.3 Manufacturing, Production, and Engineering Management

Like other departments, manufacturing also has operating objectives and strategies. Of course, these are a primary concern in the development of a manufacturing plan. It is necessary to uncover both the written and unwritten policies that detail the how, when, where, and by whom the manufacturing will be done.

While a fine level of detail is not required at this stage, it is important to document the generally accepted manufacturing policies and constraints. Priority can be placed, then, on those aspects of the manufacturing plan that will address these concerns. Specifically, any policies, constraints, or strategies regarding the following areas should be defined:

1. Labor. Some companies have a policy regarding the number of people that are considered manageable within one facility. Also, the facility itself may have a ceiling on the number of people it can house.

Objectives and Goals

2. Space. General strategy concerning the addition (or deletion) of square footage should be noted. The facility may be limited in its flexibility to add or delete square footage. This could be due to physical limitations, legal limitations, or perhaps management limitations. Also, location and expansion policy should be noted. For example, a policy may state that any manufacturing growth will be done either in an alternate facility or be off-loaded to vendors.
3. Product/process priority. Although a specific priority list is not necessary, it is important to understand which product or components of products should definitely be manufactured in the facility being planned. For example, some aspects of the products or the manufacturing process require more support from management and engineering than do other aspects. The major product/process priorities should be clarified and documented.
4. Miscellaneous. A number of other broad-based strategies, policies, and limitations may be documented at this early stage. Depending on the company, these items may include productivity assumptions, return on investment strategies, policy regarding the number of shifts, etc.

Through understanding the general manufacturing strategies, operating parameters, and limitations, the effort required to develop a manufacturing plan can be concentrated on the areas requiring the most attention. The concerns and the focus of the manufacturing plan obviously should be in harmony with the major concerns of the business, whether these are manpower, equipment, square footage, cost, etc. Everyone within the manufacturing organization maintains essentially the same general perspective on manufacturing.

Unfortunately, while their perspective on the manufacturing function is the same, the production management and production engineering people typically differ on the goals and objectives for the manufacturing plan itself. This is especially true when the manufacturing plan is being done primarily to support the efforts of manufacturing—as opposed to executive corporate—management. In this case, manufacturing planning is usually done to support the purchase and installation of new equipment, the hiring of new people, and the construction of new facilities.

However, in this situation, the interests of the manufacturing managers are different from those of the engineering managers.

The entire incentive structure for manufacturing/production management is oriented toward meeting shipping schedules while remaining within budget. It is in their interest to ensure that more than adequate manufacturing resources are available. Managing a factory with excess capacity is obviously much easier than managing one that is producing to its maximum capability.

The problem of vested interest becomes pronounced when manufacturing planning is done by production management. In many cases, especially if done as a rush job, a production manager's motivation will lead him to overestimate the resources required. Conversely, manufacturing/industrial engineering has a tendency to underestimate the resource required. These engineering groups are primarily rewarded for minimizing manufacturing costs. The manufacturing plan will be viewed as an opportunity to tighten up the manufacturing process and, perhaps, add some new improvements.

For both groups, a detailed manufacturing plan can provide guidance and justification for the acquisition of manpower, equipment, and facilities. Engineering will view this as an opportunity to justify the implementation of more productive methods and the purchase of fancy equipment. Typically, production management would rather expand on the existing operation and thereby minimize disruption and production delays. Handled properly, the manufacturing plan can be of equal benefit to both groups.

3.2.4 Time Management

While various functions have their own objectives and goals for the manufacturing plan, time is also an important consideration. The time allotted to develop a manufacturing plan is usually established by the primary user of the plan. Like any plan, a manufacturing plan must be done on time and it must be accurate. Unfortunately, these are usually conflicting goals. To resolve this conflict a compromise must be reached between the time required versus the goals desired.

The accuracy of manufacturing plans, like that of other plans, increases with the amount of time spent in developing those plans. Similar to most learning curves, the relationship is not directly proportional. For example, after reaching a

Objectives and Goals 43

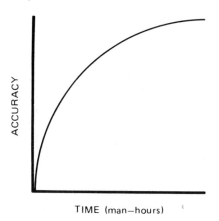

Fig. 3.1   Accuracy versus time.

certain point in time, any additional time will not significantly increase knowledge in the learning process, nor will it increase accuracy in the planning process (Fig. 3.1).

This interaction between time and accuracy can also be expressed in management's favorite terminology: dollars and cents. It is easy to recognize that the more time spent on manufacturing planning, the more that plan will cost. The costs that are difficult to quantify are those that are incurred because of an inaccurate plan or, worse, no plan at all. Ideally, manufacturing

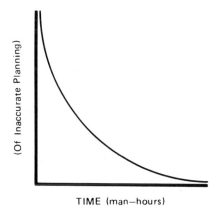

Fig. 3.2   Cost (of inaccurate planning) versus time.

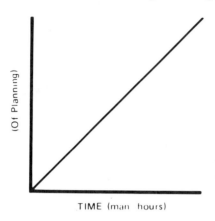

Fig. 3.3   Cost (of planning) versus time.

planning reduces costs by being reasonably accurate within a reasonable time (Figs. 3.2 to 3.4).

The primary requestors of the manufacturing plan, and their objectives for the plan, will help to determine the trade-offs involved between time and accuracy. In some plans, time is of the essence; in others, the accuracy of specific details is critical. In any case, care must be taken to define the needs of the people requesting the plan. Once defined, these needs

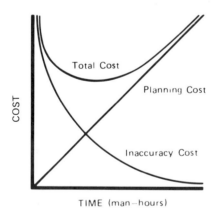

Fig. 3.4   Optimum length of time to develop a manufacturing plan.

must be communicated effectively to all persons responsible for developing the plan. Preferably, this will be done, in person, in a meeting between the planners and the requestors. Following the meeting, these needs should be documented to ensure clarification and agreement.

3.3 CENTRALIZED RESPONSIBILITY

The responsibility for coordinating, controlling, and preparing a detailed manufacturing plan should reside with one individual or, if necessary, a project team. Depending on the size of the company, the effort required could range from a full-time staff under the vice president of manufacturing to the part-time task of an industrial engineer. In any case, the development of a manufacturing plan requires a dedicated and centralized effort. The role is necessary to ensure that the plan is accurate, consistent, and auditable.

3.3.1 Consistency

Throughout the development of a manufacturing plan, there will be a number of factors, conditions, and assumptions that will have to be defined. A centralized planning effort will ensure that the same definitions are utilized throughout the manufacturing organization. A manufacturing plan in which the terminology is consistent will be more readily interpreted by management. Time can then be spent evaluating the plan rather than deciphering the plan.

Not only must the terminology be consistent in a manufacturing plan, but so should the methods used in developing that plan. Like many other endeavors, there are a number of different ways to attain the same result, in this case, a manufacturing plan. Although there is no "right" approach, the approach used should be consistent throughout each factory or manufacturing department.

While standardized definitions and methods are important to a comprehensive manufacturing plan, they do no good unless presented in an intelligible fashion. A centralized planning function can ensure that the manufacturing plans, for each production department, plant, or factory, are presented in a consistent and readily understood format.

3.3.2 Auditability

Consistent terminology, methods, and presentation formats will ensure that each aspect of the manufacturing plan can be inspected in detail. An easily audited manufacturing plan can go a long way in demonstrating its integrity and, therefore, maximizing its credibility.

The ability to audit a manufacturing plan is also necessary due to the number of judgments and assumptions that have to be made during the planning process. Any dispute with these judgments can then be can be explained, or corrected if necessary. The visibility of detailed backup data will go a long way toward selling the integrity of the plan. A centralized and organized manufacturing planning effort can more easily provide access to intelligible, organized backup data.

An independent individual or centralized project team can minimize the introduction of biased or subjective data into the manufacturing plan. An objective plan that is based on the "best" available data at the time should have nothing to hide. The integrity of the manufacturing plan can be maintained best by a centralized planning function. For example, as mentioned earlier, problems can occur if manufacturing planning is done by production management rather than a staff function. All too often it is in their vested interest to ensure that more than adequate manufacturing resources are available. However, overstating future resource requirements by production management is not necessarily done as a deliberate and deceitful strategy. Rather it occurs because of limited time and a bias that is reflected in the many assumptions and judgments that must be made during the planning process.

3.3.3. Time

A dedicated and centralized manufacturing planning function can help ensure that adequate time is devoted to developing the plan. Inadequate development time can lead to an inadequate plan as was demonstrated in the earlier discussion of time versus accuracy.

3.3.4 Maintenance

A consistent and auditable plan requires maintenance. The assumptions, base data, and calculations require organization.

Maintaining these records in an organized fashion should be the responsibility of a centralized planning function. Inquiries regarding the plan can then be readily answered and changes easily incorporated.

3.4 MANUFACTURING PLANNING AS AN ONGOING FUNCTION

There are many industries and companies where manufacturing plans, even when revised annually, are insufficient to meet the rapidly changing needs of the business. The same factors that contribute to the need for manufacturing planning may actually necessitate a full-time manufacturing planning system. A manufacturing planning system is a computer-aided or manual procedure that is in place and ready to run at any time to develop a manufacturing plan. An effective manufacturing planning system has the capability of rapidly and accurately forecasting manufacturing resource requirements at any time throughout the year based on new marketing forecasts or alternative manufacturing strategies.

Even when a one-shot manufacturing plan is requested, it is usually advantageous to develop a manufacturing planning system. There is a negligible difference between the effort required to develop a one-shot plan and the effort required to develop an entire manufacturing planning system. The data that need to be collected and utilized are the same. What is different is the organization of the data. By deciding in advance to develop a reusable system, care can be taken throughout the process to ensure that the data are properly organized.

Developing a planning system as opposed to a one-shot plan has a number of advantages. The primary advantage is that a new plan can be generated with relative ease. Regenerating a manufacturing plan is inevitable, since operating assumptions are subject to change, the marketing forecast will be revised, or an alternate strategy will be dictated by management. A planning system will allow these changes to be readily incorporated into a new plan.

A manufacturing planning system should not be developed for the sole purpose of begrudgingly handling the inevitable changes. Rather the beauty of a planning system is that assumptions, base data, and variables can be altered easily. By exploiting this capability, a more refined and well-conceived manufacturing plan can be generated.

3.4.1 Flexibility

A well-constructed manufacturing planning system is flexible. Its ability to handle alternative marketing forecasts, production factors, and manufacturing assumptions as variables is of significant benefit to the business.

An effective manufacturing planning system can be used to generate manufacturing plans that will relfect the impact of alternatives:

1. Marketing forecasts. When a single manufacturing plan is generated, only the "most likely" marketing forecast can be used. On the other hand, a manufacturing planning system can be used to evaluate the impact of a variety of forecasts from a "worst case" projection to the "best case" projection.
2. Production strategies. Although the marketing forecast may remain constant, the strategy for producing to that forecast may not. For example, management may wish to consider the impact of making of buying certain products and subassemblies. Many companies have the option of utilizing alternative factories, overseas subsidiaries, or outside vendors for the procurement of some of the subassemblies or products. Production strategies also include overtime, second/third shifts, and other in-house manufacturing alternatives.
3. Productivity plans. A manufacturing planning system should be an effective tool for evaluating the impact of proposed changes to the production process. For instance, a planning system could quantify the impact of improving equipment uptime or replacing the equipment, to reducing scrap or rework. Assessing the impact of improved attendance to reducing labor hours could also be done.

3.4.2 Multiple Users

A manufacturing planning system can be used by a number of departments, such as finance, marketing, product design, production control, production management, and industrial engineering, each with its own purpose:

Manufacturing Planning as an Ongoing Function

1. Finance. Financial planning and budgeting for the manufacturing operation are simplified and made more accurate through the use of a manufacturing plan. Rather than using trend analysis to extrapolate future capital requirements, a manufacturing plan can specify in detail the resources to be acquired. Combining a resource forecast with the cost of each resource can provide a more detailed and accurate financial plan. A manufacturing planning system can also assist the financial department in analyzing the validity of a capital appropriations request, help quantify the benefits of a productivity improvement project.
2. Marketing. The marketing group may wish to review the feasibility of alternative marketing forecasts. For example, the marketing department might wish to trim down an optimistic forecast after realizing the impact of that forecast on the manufacturing operation. A planning system can provide a rapid response to a marketing forecast. If needed, the forecast can then be synchronized with manufacturing capacity. Although manufacturing should not dictate the marketing forecast, it is important that marketing know as soon as possible if the forecast will be impossible to produce.
3. Product design. A manufacturing planning system may also be utilized by product design engineers to evaluate the producibility and impact on manufacturing of alternative product designs. Not only can the alternative product designs be evaluated individually, but the planning system can put these designs in perspective. For example, in the context of the entire marketing forecast, an alternative design of one product may have an insignificant impact on the production operation.
4. Production control. A long-range manufacturing planning system may also assist the production control department even though its concern is with short-range scheduling. For instance, production control may wish to utilize the planning system to assess lot sizing strategy for subassembly manufacturing.
5. Production management. Production management will be one of the more significant beneficiaries of an effective manufacturing planning system. Production management can use the system frequently and for a number of purposes such as:

Manpower planning. A planning system will eliminate the need for production management to manually calculate future production labor requirements. Once these managers are convinced of the accuracy of the manufacturing planning system, they can rely on it for their own manpower planning.

Shift strategy. Management can review the consequences of changing the number of shifts that a work center or a whole department will operate. For example, rather than buy new production equipment, a manager may opt to run that department with three shifts rather than two.

Variance justification. A manufacturing plan may help prove that a cost variance has been incurred by a product mix that is different than that used in the financial plan.

Capital appropriations. Production management may wish to document the need for a capital expense by attaching a relevant portion of the manufacturing plan.

6. Industrial engineering. As a manufacturing service organization, industrial engineering should have as many or perhaps more uses for the manufacturing planning system than production management. In addition to the same uses as production management, industrial engineering may wish to utilize the system to conduct studies in a variety of areas such as:

Make of buy decisions. A planning system will expedite the development of manufacturing costs for subassemblies and products. Also, manufacturing costs can be developed within the context of projected capacity. For instance, manufacturing the product, or subassembly, could require the acquisition of additional capacity or simply make use of existing capacity.

Productivity planning. Industrial engineering may wish to play out alternative scenarios which will reflect the impact of various productivity improvements. The merits of those improvements in relationship to projected production rates can be evaluated and quantified.

Manufacturing Planning as an Ongoing Function 51

> Plant layout. A planning system can rapidly provide the data necessary to develop both "rough-cut" block plans and detailed one-quarter-inch-scaled layouts.
>
> Operational studies. A planning system can be used to identify and resolve a variety of operational problems. A system can be used to highlight problems regarding labor efficiency, equipment uptime, or space utilization. The system may also be used to evaluate the impact of corrective actions.

3.4.3 Cost Effectiveness

The expense required to develop and maintain an ongoing manufacturing planning system is small compared with the cost that may occur without such a system. The cost of maintaining a planning system is fixed, while the potential cost of ineffective planning is unlimited.

The benefits derived from a one-shot manufacturing plan were highlighted earlier. A planning system will provide the same benefits but to a much greater extent. The benefits of a manufacturing plan are greatly increased with a computer-aided system's ability to readily regenerate new plans and therefore the number of applications for which the system can be used. This capability renders it nearly impossible to spend too much money on a manufacturing planning system.

Of course, the effectiveness of a planning system depends not only upon its accuracy, but that it is used. A well-conceived planning system that inspires confidence will be used. The ability to be used and reused for a number of purposes will result in additional cost benefits not found in a one-shot manufacturing plan.

The same chart (Fig. 3.4) presented earlier, which demonstrates the relationship of man-hours and accuracy to cost, can be used to illustrate the effectiveness of a planning system. With a one-shot manufacturing plan, additional man-hours beyond a certain point may not contribute to any significant additional accuracy or benefit. With a planning system, the same is true, except there is significantly less time/cost needed to generate a plan with the same level of accuracy.

There are not guidelines that can be used to establish specifically what a planning system should cost to develop and maintain. Obviously it is contingent on the size and complexity of the manufacturing operation. It is also contingent on the accessibility of pertinent data. The planning systems themselves can range from a rough-cut to a highly detailed system. With this disclaimer in mind, it should be reemphasized that the development of a planning system should not take much longer than the development of a comparably detailed manufacturing plan. What does take additional time is the maintenance of a planning system. Fortunately, the additional maintenance effort is more than offset by the capabilities of a manufacturing planning system.

It is impossible to establish a specific guideline for the number of man-hours that should be committed to a manufacturing planning system. It is, however, possible to put this effort in perspective and note a specific example. A division of a major high-technology company has required the full-time efforts of an industrial engineer plus additional programming assistance from time to time. The division employs roughly 1000 direct labor workers in a highly sophisticated and rapidly changing environment. Correspondingly, the system in place is also highly sophisticated, especially since it must keep up with the rapid changes in both the product and the process. After 3 years of full-time support, a computer-aided planning system is still being refined and its coverage extended. The system is used to generate roughly 60 manufacturing plans per year. The system can handle only one forecasted time period at a time. It is used to evaluate marketing forecasts for a variety of time frames and also to evaluate alternative strategies and operating conditions. In a more traditional manufacturing environment, a planning system should require much less effort. A good example of this can be seen in a consumer goods plant. Here there are four major high-volume products being produced by 1500 workers. The production process as well as the product lines are also relatively static, showing slow growth. In this environment, the system still may require 6 to 9 months to develop, but once developed the system may require maintenance only 1 day/month.

Manufacturing planning, as an ongoing responsibility of an individual or department, will easily justify the commitment it deserves.

3.5 ORGANIZATION

The manufacturing planning process must be organized. As discussed earlier, the planning effort must be centralized. In most

Organization

cases, this will mean assigning one individual as a coordinator. As a coordinator, his efforts are not autonomous. Cooperation and constructive effort are required from others throughout the organization. This organizational contribution must also be managed properly. Like any well-managed project, a project plan will help organize the manufacturing planning effort.

3.5.1 The Coordinator

The person responsible for coordinating, developing, and generating a manufacturing plan is obviously key to the success of that plan. A wide range of knowledge and abilities along with some personality traits will help to ensure a coordinator's effectiveness. A manufacturing planner must have general product and process knowledge. This includes a general understanding of the product line, including its subassemblies and components. Also needed is an understanding of the manufacturing methods used in producing each of these products, and a feel for how the factory is managed. Beyond knowing who the players are, an understanding of the factory control systems and their effectiveness may be helpful. For example, knowing that the labor reporting system is inaccurate could be essential.

Depending on the plans objectives, a facilities background in plant layout may be needed. If the focus of the manufacturing plan is on facilities and space, one must consider the constraints and limitations of both the facility and the process. Knowing these limitations could eliminate from further consideration some plant layout or expansion alternatives.

Depending on the scope of the manufacturing plan or planning system, it may be necessary for the manufacturing planner to be computer literate. The planner does not have to be a computer programmer, but rather an understanding of a computer's capabilities may be beneficial. Some computer knowledge may enable the planner to design a computer-aided manufacturing planning system and draft a specification that can then be communicated to a programmer. Even when developing a one-shot plan, some of the necessary data may be on a computer file. Interpreting these data may be greatly simplified by accessing only that which is needed. At least a rudimentary knowledge of computer systems will enable the planner to know what is and is not possible.

The planner must also have the ability to perceive the "big picture," as well as pay meticulous attention to details. Details and numerous repetitive simple calculations can be taxing.

Because of the volume of data and the number of calculations, this process requires an organized individual.

Communication skills can be a major asset to the manufacturing planner. It is inevitable that both written and verbal communications will be necessary. The planner must gather information from upper, middle, and line management. These same managers most likely will expect both written and verbal presentations of the plan's results. When collecting information, the planner may have to act as a negotiator. Each manager may request that his or her own operating assumptions be used in developing the plan. If the assumptions suggested by each manager are in conflict, a compromise must be worked out.

Finally, the planner must be confident and objective. Every effort must be made to ensure that the correct assumptions are used and that the right decisions are made. The planner must be diplomatic yet assertive and not be pushed into using questionable data.

3.5.2 Organization

In most companies the responsibility for manufacturing planning is or should be a function of the industrial engineering department. In large companies there may be sufficient work to warrant a separate planning group staffed by industrial engineers. The knowledge and skills needed for manufacturing planning are industrial engineering skills. The industrial engineering department also deals with most of the raw data that are necessary to construct the manufacturing plan. When pertinent production data are not immediately available, the industrial engineering department should know where to get it. Industrial engineers are well qualified for manufacturing planning because much of the industrial engineer's role involves quantifying the relationship between the product and the process.

In some cases, it may be necessary for a company to hire a management consultant. An experienced consultant in manufacturing planning can provide the expertise that is not available in-houses. The consultant may also ensure that the plan is done on time, the deadline met. In other instances, an objective opinion may be another reason to utilize a consultant. It is important to note that although a consultant may be an expert, a planner/coordinator is still necessary. The coordinator can provide the interface between the consultant and the company.

Organization

The consultant will know what information is needed and the company's liaison will know where to get it. The coordinator will know what documents are available and which people to talk to.

Coordination and cooperation with other departments throughout the company is essential to the manufacturing plan. Whether done by a consultant or an in-house planner, assistance may be required from nearly every department within the company. Assistance can range from providing a simple answer to contributing a major effort. Obtaining the cooperation of other departments and avoiding problems will make the planning process much easier. Gaining the support of others is somewhat a matter of style and a method of approach, as opposed to a specific technique. Being both honest and open is most effective in the long run.

To foster a cooperative environment, a memo should be drafted, preferably by upper management. This memo should express a commitment by upper management to the manufacturing planning effort. It should also contain a brief summary of the objectives that are to be achieved through the manufacturing planning. Perhaps, most important, the memo should introduce the manufacturing planner. The planner's role or function within the planning process must be communicated. The memo should then be sent to every manager from whom the planner requires information or assistance. The production managers and various production engineering managers would be of primary importance. It may also be necessary to contact the managers of cost accounting, product design, marketing, and production control. These managers are usually exceptionally busy. They are more likely to cooperate and support a manufacturing planner if they understand the importance being placed on manufacturing planning and the planner's role. In the interest of maintaining an open and honest dialog, these managers should be brought into the planning process as early as possible. Keeping them informed and, hopefully, getting them involved will help to foster a cooperative environment. As an insider to the planning process, a manager's contributions are more likely to be significant in terms of both quantity and quality.

It is important to recognize that communication is a two-way street. The planner should share and allow others an opportunity to review all of the base data, especially the assumptions that are contributed by various managers and others. These managers should also have access to the data and assumptions developed by the planner. If these managers contribute or concur with the building blocks of the plan, they are more likely to support the outcome of that plan.

3.5.3 The Project Plan

Developing a manufacturing plan or planning system is a project. Like any management or engineering project, the documentation of a project plan may be helpful. The project plan should help to ensure that nothing is overlooked, that a time table is kept, and that the project's information and time requirements are communicated.

The first two steps of any manufacturing plan have already been discussed. The first is to define the objectives of the manufacturing plan in the context of the overall business and corporate objectives. This step will provide the direction for the planning process. It should answer the questions "Why are we developing a manufacturing plan?" and "What do we expect to get out of a manufacturing plan?". The second step is to organize a planning effort. The question answered here is "Who is developing the manufacturing plan?". In this step an individual is designated as the manufacturing planner and his or her role is defined. Likewise, the role of others in the planning effort is also defined.

Once the objectives are understood and the effort is organized, it then becomes necessary to define a procedure, in other words, "How is manufacturing planning done?". Answering this question in detail is the thrust of the rest of this text. Even so, at this point it is important to understand each step in the context of the overall procedure. The procedure for developing a manufacturing plan is outlined below:

1. Collect product data: the marketing forecast; secondary products; new products; product design changes; product configuration
2. Collect manufacturing process data: manufacturing operations in sequence, labor planning data; equipment planning data
3. Collect facilities planning data: equipment square footage; work-in-process; indirect areas; warehousing and distribution; administrative and technical offices
4. Calculate manufacturing requirements
5. Develop a plan for meeting those asset requirements

Probably the most difficult aspect of developing a project plan is to estimate time, and answer the question "When will each step be completed?". It is impossible to provide a rule of thumb for estimating the time required to develop a manufacturing plan. It is, however, possible to suggest that the data collection phase is by far the most time consuming.

The work involved in each step of the planning process will be presented in detail throughout the text. Only through understanding the objectives desired and the work required can a realistic schedule be attached to the planning procedure. However, it is highly unlikely that any planner will have the luxury of developing one's own implementation schedule. It is more likely that a deadline will be dictated. A schedule will have to be constructed to fit this deadline. The deadline will determine the level of accuracy that can be achieved. It will probably not change the percentage of time that will be spent on each step. For example, probably 70% of the time allocated will be spent simply in collecting and organizing the data. The nature of the industry and the objectives of the plan will determine the breakdown of that 70%. Only 10% of the time will actually be spent number crunching the manufacturing requirements. Another 10% of the time should be allocated to evaluate alternative scenarios for meeting those requirements. This will leave 10% of the time available as a contingency and it probably will be needed. It is inevitable that during the planning process new data will emerge or be dictated that will affect some of the basic assumptions being used. Also, disagreements may ensue regarding these assumptions which will contribute to slowing the planning process. Preparing for this situation will minimize the possibility of missing one's deadline.

REFERENCES

1. W. Ouchi, *Theory Z*, Addison-Wesley, Reading, Mass., 1981.
2. B. Shapiro, Can marketing and manufacturing coexist?, *Harvard Business Review*, September/October 1977.

4
Data Collection

4.1 IMPORTANCE OF DATA

Along with being the most time-consuming portion of manufacturing planning, data collection is also the most important. Each piece of data is an integral part of the foundation of the manufacturing plan and must be carefully defined. Once defined, its use must also be careful and consistent throughout the planning process. A good example of this need for data elements and terminology to be defined will be seen later in the development of production factors such as attendance. The definition of attendance can vary greatly. A number of factors can be included or excluded when calculating attendance. Specifically, these factors include vacation, holidays, training time, paid and unpaid sick time, etc. The point is that therminology must be specifically and consistently defined in order to produce a reliable manufacturing plan.

In other cases, the data being collected will be predefined by the company or industry. It is still important that the accepted definition be understood and, if possible, documented.

Documentation will help to ensure consistency and also to ensure that no oversights are made. Oversights regarding a clear definition of the data could lead to inaccuracies and confusion, and therefore to a lack of credibility for the manufacturing plan.
The planner should avoid making assumptions but rather should ask what is or is not included in each fact or figure. Clarification will help minimize the possibility of double counting or omitting key components of the manufacturing planning process.

Collecting the data can begin once it is understood what data are required and how they are defined. Unfortunately, getting the data is not usually as simple as it sounds. The data may not be easily accessible in the format one desires; it may be inconsistent or, worse, nonexistent. Due to the volume of data required for manufacturing planning, problems in its collection is inevitable. Some of these problems are predictable and therefore can be avoided or at least their impact minimized. Avoiding these problems requires knowing what data are needed, how they are to be defined, and where they can be found. The data collection process should be well conceived and organized.

4.2 DIFFICULTY OF COLLECTING DATA

4.2.1 Timeliness Versus Accuracy

Time/cost versus accuracy is a familiar problem that has already been discussed in the context of the overall manufacturing plan. The same conflict also occurs during the data collection process. Not every piece of data is clear-cut as to its meaning or definition. Knowing when, and when not, to take the time to investigate can be critical.

A judgment must be made regarding the piece of data in question. Its significance to the outcome of the manufacturing plan must be determined. In most cases, this can be done by gut feel rather than a detailed study.

Unless the planner is careful, it is very easy to get bogged down overanalyzing one small piece of the manufacturing plan. Every once in a while, it is necessary to step back and place these decisions in the context of the overall plan. If this perspective is maintained, then the effective utilization of time will follow. A proper perspective would include maintaining a constant awareness of the objectives of the plan. For example, assume that it would take a major effort to check and refine the

Difficulty of Collecting Data 61

accuracy of a piece of data. Assume also that that data would have a significant impact on space requirements. If the objective of the manufacturing plan is to allocate space, then these data must be investigated. On the other hand, if the primary objective of the plan is to determine manpower requirements, then refining these data could wait until later.

The time required to investigate a piece of data also is a consideration. In some cases, it might be necessary to clarify the definition of a piece of data through an investigation rather than estimate its impact on the manufacturing plan. At other times, an estimate of the data's validity might suffice, especially if the time required for investigation is prohibitive.

By maintaining one's perspective, the tendency to collect too much detail can be avoided. Usually, collecting too much detail is a waste of time. An overemphasis on detail also can narrow one's perspective by focusing too much attention on one small aspect of the plan. Avoiding too much detail is easier said than done. Many people are more comfortable discussing the fine points rather than the "big picture." This fixation on detail leads to a situation where detailed information is collected on everything pertaining to the manufacturing operation. The tendency to collect too much data should be resisted. Only the appropriate level of detail that is required for planning should be collected. If necessary, it is easy to go back and collect any additional data that may be required.

4.2.2 Data Are Not Constant

Both the product and the process are constantly in a state of change. Monitoring these changes requires a multitude of facts and figures. Some of these figures require routine maintenance, while others require intermittent adjustment. Many constantly changing facts and figures are required to develop a manufacturing plan. Data that are not fixed pose a problem for the manufacturing planner. It may appear impossible to keep current with, let alone use, the data in a manufacutring plan.

Fortunately, most of these facts and figures, such as industrial engineering standards, rework rates, equipment uptimes, or product configuration ratios, do not change drastically from month to month. However, to minimize confusion, a date should be established for planning purposes. The plan can then be based on the manufacturing operation as of that date. The foundation of the manufacturing plan can then be stated simply: "Based on the way we operate today, these are the resources that we will need in the future."

In addition to the data that are maintained routinely, there are also other changes to the product and process to be considered. For example, it may be stated that the company is going with a new process or a new product design next month or next year. The planner must decide which changes to incorporate into the manufacturing plan and which changes to ignore. A guideline established at the beginning of the planning process may eliminate problems later. A guideline should state at what point a change will be considered in the manufacturing plan and at what point it is considered too intangible to be considered. Each company should establish its own guidelines based on its ability to predict these changes. A good example can be seen when changes are being made to the manufacturing process. A guideline might state that the change should be incorporated in the plan if the equipment has already been ordered, if the plant layout has been signed off, and the whole project is approaching implementation. Conversely, a guideline might state that if a process change is still in the conceptual stage it will not be considered in the manufacturing plan. For example, the following statement should be considered conceptual: "We are looking into robots for that area by next year."

In manufacturing planning, the base data are somewhat like a picture of a moving target: its an accurate portrayal of only a single moment. In reality, the target does not stand still. In a company, there are always organizational shakeups and new systems and procedures being installed. At some point, a limit must be established as to what is to be included in the picture. By giving this some thought and establishing guidelines, confusion can be avoided and the productive planning time maximized.

4.2.3 Contradictory Information

Collecting data for a manufacturing plan will require a wide variety of sources. It is inevitable that at some time data collected from two sources on the same subject will be in conflict. This is frustrating, but steps can be taken to handle this situation. Essentially, some of the same ground rules apply here that applied to the timeliness versus accuracy dilemma. First the

Difficulty of Collecting Data

conflict must be put in perspective. If the conflict is insignificant to the overall manufacturing plan and its objectives, then there is no need to resolve it. Either piece of data could be used and supply essentially the same answer.

The conflict must be resolved when its impact is significant to the manufacturing plan. Again the conflict must be resolved in the context of the manufacturing plan. Too many times conflicting data are the result of conflicting objectives. A good example can be seen in the determination of the absenteeism. The cost accounting/payroll department will probably define the absence rate quite differently from that of the production manager. The primary objective of the cost accounting department is to monitor and control costs. Therefore, their definition of an absence percentage may be based on dividing the dollars paid for absent hours by the total dollars paid. While this definition may meet the objectives of financial planning, it is not well suited for manufacturing planning where consideration of unpaid absence is also necessary. The production department depends on an adequate work force. Whether those absent are paid, or unpaid, is of little significance to the management of the production operation. While both definitions are correct in the context of their objectives, only one is well suited to manufacturing planning.

Once in a while it may be necessary to consult upper management on the resolution of a conflict. A decision can then be rendered on which piece of data should be used for planning purposes. A typical example can be seen in determining which manufacturing method to use for planning purposes. The production manager's definition of how the process has been working may be quite different than the manufacturing engineering manager's definition of how the process should be working. In this case, it might be necessary to consult the vice president of manufacturing and then proceed on the basis of his or her decision.

4.2.4 Documentation

Perhaps more important than the data is that the data be documented. Notes on how the data were generated and defined, and assumptions made are essential.

The accuracy of the plan depends on the accuracy of the data. It also depends on consistent definition and utilization of the data throughout the plan. Developing a manufacturing plan requires processing a large volume of data. Managing these data and ensuring their consistent and accurate application will require that the foundation of the data be recorded. Even if the planner is unsure of the accuracy of the data, documentation of that fact can actually enhance the plan's credibility. No one can expect a projection of the future to be done without making some judgments. What will be expected is a record of how and why these judgments were made. As long as this information is available, it will go a long way toward instilling confidence in the plan and minimizing disputes with it.

By adequately documenting the basis of the plan, the plan becomes indisputable in a sense. If one agrees with the foundation of the plan and the planning process itself, then one should agree with its outcome. Conversely, if one disagrees with the outcome of the manufacturing plan one should also disagree with some portion of the planning process or its foundation. It is essential that both the data base and the calculations used be adequately documented. The ability to audit the manufacturing plan may also be of significant benefit in the future. A manufacturing plan may be used months or perhaps years after it is prepared. At a future time a plan may be useless unless its foundation has been sufficiently documented.

Thorough documentation is also helpful when a long-range plan is published and utilized by other departments. In order to utilize the plan effectively for their own purposes, certain aspects of the plan may have to be enhanced. A close inspection of the backup documentation, assuming it is available, may be necessary. In most cases a manufacturing plan will have to be revised after it is first generated and presented. Unfortunately, but typically, revising the plan may require altering what was considered an unalterable piece of data. Only by knowing what these data were, and how and where they were used, can the manufacturing plan be accurately revised.

4.2.5 Organizing the Data

Accurate documentation will enhance a manufacturing plan's credibility, auditability, and flexibility. Of course, these benefits depend on the documentation being readily accessible and presentable. Care must be taken to organize documentation in a logical and coherent fashion.

The ability to generate a manufacturing plan also depends on the data being properly documented and organized. Primarily, this dependency on organization is due to the volume of data and the time required to collect them. Since the data may be utilized well after they were collected, proper organization now can streamline the planning process later.

A manufacturing plan is the result of combining information about the company's products and the manufacturing process. Files should be constructed within these groupings. For example, under the heading of product data there may be a separate file constructed for each of the company's major products. Likewise, the process category may require a separate file for each of the company's manufacturing departments. While this example may be oversimplified, the content of this text will provide the detail necessary to develop an orderly system for maintaining records.

4.3 VARIABLE VERSUS FIXED DATA

When collecting and organizing data, a distinction should made made between the data that are considered fixed and the data that are considered variable for planning purposes. Knowing this distinction in advance may dictate how the information is collected and organized. This distinction can and should usually be made at the onset of the planning process. For example, in a manufacturing plan that is to look over each of the next 5 years, the marketing forecast will be the primary variable. The secondary variable might include incorporating factors for productivity improvement. Similarly, alternate manufacturing strategies may be considered a secondary variable. Secondary variables might include such factors as the number of shifts and process flow alternatives.

In most cases, much of the data that are used for manufacturing planning are essentially fixed. That is, once the data are collected, they can be used for all aspects of the manufacturing plan. For most manufacturers, the product construction and manufacturing process are essentially static. For any given manufacturing plan, the latest data available can be used for the whole plan. While processing times and bills of materials are subject to change, they do not usually change as a variable to the manufacturing plan.

Depending on the nature of the industry and the objectives of the manufacturing plan, there may be a grey area between what is considered fixed data and what will be considered a variable input to the manufacturing plan. It should be kept in mind during the data collection phase to categorize data as either variable or fixed.

4.4 SUMMARY

Collecting and organizing data are both the most important and the most time-consuming aspects of developing the manufacturing plan; however, there are some guidelines that can be followed to make this process a more productive endeavor:

1. Avoid overemphasis on detail. It is not worthwhile to make a major commitment in time to make a minor improvement in accuracy. Also, sometimes it is easier to go back for more information, rather than collect too much initially.
2. Take a picture and go with it. The product and the process are constantly in a state of change. Collecting information requires snapping a picture of a moving target. Although flexible enough to be altered, this picture will represent the foundation of the manufacturing plan.
3. Err on the high side. It is better to slightly overstate rather than understate future manufacturing requirements. Usually the loss of shipments and, therefore, business and market share is much more destructive than maintaining a slight excess of capacity. When in doubt, always play it safe—overestimate.
4. Enlist support and cooperation. The data collection phase is the time to enlist confidence, understanding, and assistance. Gaining support and cooperation early will enhance accuracy and also a sense of cooperative commitment. Also, it will help minimize any future resistance.
5. Organize and document the data. This is most important, since no one can dispute the plan if the backup data are made public: One will have to dispute with a portion of the backup data rather than the plan itself. Attention will then focus on specific problems rather than on intangible feelings.

5
Product Data

5.1 INTRODUCTION

A manufacturing plan is the result of combining knowledge of the product and the process. Therefore, the product must be defined in the context of manufacturing, as well as the overall business. The collecting, documenting, and organizing of the product data for manufacturing planning is essentially a four-step process:

1. The end products of the company must be defined. Usually, the end products will be presented in a marketing or sales forecast. In addition, consideration must be given to secondary products that are marketed to the customer. Secondary products would include such items as spare components or add-on options.
2. The products that are produced by each factory or manufacturing department must be established and documented. For instance, in a factory that assembles only components of the company's end product, those components are the products of that factory.

3. The factory's component forecast must be extrapolated from the company's marketing forecast. In some industries, the conversion is not a simple procedure.
4. For long-range planning, it may be necessary to incorporate projected product design changes and the impact of new products technologies.

5.2 THE COMPANY'S PRODUCTS

To develop a comprehensive manufacturing plan, it is imperative that one knows every product that the company or divisions manufactures and markets. It is also important to understand the extent to which all these products are incorporated in the marketing forecast. Secondary products that are omitted in the marketing forecast must still be factored into the manufacturing plan.

5.2.1 Marketing Forecast

For most manufacturing plans, the marketing forecast is the primary variable. It is the primary determinant of future manpower, equipment, and facilities requirements.

The manufacturing plan is only as good as the marketing forecast on which it is based. A detailed knowledge of what is and is not included in that forecast is essential. The accuracy of the manufacturing plan depends on it—not to mention the future of the business.

Although critical to the manufacturing plan, developing a marketing forecast is beyond the scope of a manufacturing planner. However, in some cases, the manufacturing planner can have a positive influence on the development of the forecast. Hill established four guidelines when obtaining a marketing forecast [1]:

1. "Do not let forecasters be placed in a position where they can be hung with their own forecast." An environment should be fostered which will enable a forecaster to express his or her most candid and realistic projection of product demand. Hill states, "A tendency to blame the forecaster when forecasts are not met is

The Company's Products

likely to result in conservatism rather than the realism that is most desirable."
2. "Do not ask for levels of detail that are not required for the exercise." Forecasting product demand, especially long range, is difficult enough without a fine level of detail. For long-range planning, it may be sufficient to utilize a forecast for general product groups rather than for specific products. Care must be taken to ensure that these groupings are applicable to manufacturing planning and familiar to the forecaster.
3. "Obtain forecasts from those who are most responsible for the results being predicted." The marketing and sales people are responsible for monitoring the market and achieving sales. Through their intimate knowledge of the market place, the forecast will then be based on the best information available. Perhaps more importantly, this forecast is very likely to be their goal. The forecast is more likely to become a reality if it is developed by those with the power and the incentive to meet that forecast.
4. "In dealing with sales forecasts, try to avoid obtaining forecasts merely in dollar terms." In many companies, long-range marketing projections are quantified in terms of market share and revenue. Due to the impact of product mix on manufacturing, it is imperative that revenue projections be converted into product volumes. Preferably, this conversion should be provided by the marketing/sales organization, rather than a crude interpretation by the manufacturing planner.

Once a marketing forecast is developed to the best of the marketing organization's ability, it is up to the manufacturing planner to utilize that data properly. The planner must understand the forecast and the assumptions and intentions that went into it.

In most instances a marketing forecast will be presented on one sheet of paper. There will be a list of each major product of the company and a corresponding production volume for each of the time periods being forecasted. Chances are the forecast will not show whether the product should be considered a stripped down version or a fully loaded version. If it is based on the current typical version, this fact should be understood by the manufacturing planner. Also, the planner should know which products are subject to change during the planning process. For example, in generating a marketing forecast, the forecaster may estimate a product volume to meet a deadline. However, on further investigation, the projection may become refined and the manufacturing plan altered to reflect that change. Knowing this in advance will enable the planner to more easily accommodate these changes.

The marketing forecast should project a production volume for all of the company's major end products. Sometimes this forecast might include a volume for each specific product. At other times, it could be a volume for product groupings. In any case, the bulk of what the company ships should be in the marketing forecast. Once it is understood what will be forecasted, an effort should be made to uncover all of those products that are not forecasted. These secondary products may be just as vital to certain portions of the manufacturing process as the company's end products.

5.2.2 Secondary Products

In addition to its major products, it is important to consider anything else that the company ships. Secondary products would include such items as spare components, second-source components, or product enhancement options. Anything that contributes an additional work load on the factory must be considered.

Essentially, there are two ways to incorporate secondary products in the planning process. One way is to request or establish a forecast the same way that the primary products are forecasted. If this is impossible, it may be necessary to determine a ratio of secondary to primary product volumes. Most likely a combination of both approaches will be necessary to develop the whole picture.

Spare Parts

In an assembly-oriented industry, spare parts may have a significant impact on manufacturing requirements. In most cases, spare parts are not the predominant concern of the sales/marketing organization. Therefore, spares are not usually forecasted

The Company's Products

as a part of the marketing forecast. Other means must be used to project spare parts requirements.

Although unavailable from marketing, a spares forecast may be provided from another group within the company. It may be necessary to consult with one or more of the following groups: production control, field service, the repair department, or quality control. Even if a spares forecast is not provided, their information will enable one to be extrapolated. For instance, by evaluating the past and present spares business, a projection of the future spares business can be made. A spares forecast can be extrapolated from historical data. One of the above organizations should be able to provide data on actual spares usage. Most likely this would be the field service or repair department.

Another approach would be to ratio the spare parts production rate to the product production rate. The problem with this approach is that there is usually a delay between when a product is shipped and when it will need spare parts. However, the time shift can be taken into account. For example, assume that 25% of the product shipped requires the same new component within a year. That component's spare forecast would then be 25% of the product forecast from the previous year. In industries where the spares forecast is less significant, this time phasing may be unnecessary. In other words, if spares account for only a minute portion of the total manufacturing requirements, then a straight ratio of current spares production to primary production may suffice.

Expansion Options

Other secondary products—perhaps of significant consequence—are add-on items, options, or expansions for previously shipped products. Similar to spares, these products can either be forecasted, trended, or ratioed. One word of caution, however: using the ratio technique for add-ons has the same time phase problem as is encountered with spares. That is, add-ons to a product may not be purchased for months or even years after the initial product is purchased. Computers and some industrial machines are examples of products that can be upgraded.

Data on expansions activity should come from marketing, although it may be necessary to consult production control or examine the history of shipments.

Second-Source Requirements

Second-source components or -products are those items that are being manufactured in-house, but not in support of the factory's end product. A good example can be seen when a manufacturer takes on additional outside work to utilize its excess capacity. For example, a home appliance division of a company may manufacture motors for its industrial division. These extra motors utilize the plants' capacity but have no relation to the division's home appliance products.

Second-source requirements can be difficult to predict. In many cases, they are only taken on as filler to utilize excess capacity. In that sense, second source is a strategy option rather than a fixed requirement. It may be necessary to consult with upper manufacturing management on the inclusion of second-source requirements in the manufacturing plan.

Assuming that second-source requirements are to be included in the manufacturing plan, the projected requirements must usually come from a forecast or trend analysis. The ratio technique will have no real validity since there is no correlation between the second-source requirements and the company's primary product line. However, a ratio technique might be useful in companies where secondary product volumes are minimal. A ratio technique may be employed with the understanding that its foundation may be questionable. The planner may have to resort to the ratio approach when secondary product projections are unavailable. An analysis can be made of the secondary products that have been and are currently produced. A ratio can then be established by comparing the total volume of product against the volume produced as a secondary product. This ratio will allow one to speculate on future secondary product requirements based on current and past production rates.

This ratio technique has its merits even though it is questionable. It is based solely on history of the manufacturing operation and does not reflect the true relationship between the secondary and primary products. Even so, this ratio approach may be a useful tool for "ball parking" secondary product requirements. It is much better than not factoring them in at all.

In order to make this or any ratio technique more effective, consideration should be given to the future advance or decline of the ratios. For example, it may be found that secondary products made up 5% of the total production requirements 5 years ago. Today, it may be found that secondary products account for 20% of the production requirements. Consequently,

The Products of the Factory or Manufacturing Department 73

it may be necessary for a 5-year manufacturing plan to reflect continued expansion of the secondary product business.

Secondary products are legitimate concerns and cannot be overlooked. Whether forecasted or factored, these products must be counted.

5.3 THE PRODUCTS OF THE FACTORY OR MANUFACTURING DEPARTMENT

Each of the company's manufacturing departments produces products. For most departments, their products will not be the same as the company's products. Each department is concerned only with what it produces and what it produces may be only a component for the company's primary products.

In the computer industry, the forecasted volume and type of computers are meaningless to the circuit board factory where labor standards and equipment capacities are in terms of circuit boards. Also, circuit board production rates may be impacted by the secondary requirements described earlier.

The products produced by each of the manufacturing departments must be established and documented. The intent is to understand each department's products and terminology; it is not to convert the marketing forecast to component requirements. That will be done later. At this point, it is important to know what is being produced by each factory/manufacturing department and how it is described.

The factory's products must be evaluated from two standpoints. First, it is necessary to define the items produced as perceived by each manufacturing department's management. Most likely they will perceive the product in general terms such as are used in status reports or routine production meetings. Second, the factory's products must be defined in specific terms. Typically, this will be in terms of part numbers as defined in process routing summaries or engineering standards data or production control systems.

5.3.1 General Units of Production

Each manufacturing department has its own products. These products are often referred to in general terms by production management and others. The majority of discussions that involve the plant's production rate or capacity do not refer to specific part numbers. More likely, there is an accepted and general definition of the products being produced.

Typically, products will be generalized or categorized into groups where a department produces a large number of different products. A department that produces 400 cable or cable harnesses is unlikely to reference each part number when discussing, documenting, or presenting production data. The cable types will most likely be grouped into half a dozen or a dozen categories for ease of interpretation. These groupings or general units of production for each department must be understood by the manufacturing planner. The main reason is to facilitate communication with production management and upper management. It will be necessary for collecting information and presenting information to utilize their terminology. Probably the best way to document these product groups is to consult with each department manager. If asked about the department's production rate or production capacity, the response will most likely contain a definition of that department's product groups, although to get a thorough answer some prodding may be necessary. Take the production manager of the cables department, for example. If asked what the production rate is he or she might say "x number of cables per week." Although true, this statement may be so general as to be useless. Perhaps a better way to capture the data is to ask, "What is your department's production capacity?" The manager's response might then be "It depends on the mix of cables coming through." The discussion can then lead to the product groupings on which the planner can base future communications and presentations.

These general units of production or product groups may also be clearly defined in a number of management reports. For example, a quality control study may document the reject rate by product type. Similarly, production reports may highlight production rates for the same product groupings. Capacity studies may be another source for this information. A serious capacity study will most likely document the mix of products on which it is based. Its definition of that mix will most likely be in terms of product groups.

5.3.2 Specific Production Units

While general product groups are needed for communication, detailed and specific product designations are needed for analysis.

Most of the data that are up to date and detailed enough for manufacturing planning will be found at the part number or assembly number level. Data from engineering documentation as well as some production and financial control systems are typically very specific.

In order to develop detailed, accurate, and auditable manufacturing plans, a list of all the fabrications or assemblies by part number should be collected for each department. This effort will also ensure that none of a production department's production requirements is overlooked.

The most important concern is that the part numbers must correspond with the part numbers as used in the industrial engineering standards data or the labor reporting systems, although the list of part numbers should come from a production schedule. In most companies, the production control department must supply each production manager with a production schedule at the part number level. A copy of this document should be obtained for each department scheduled. Usually, these schedules will, in fact, correspond to the industrial engineering standards and labor reporting numbers. However, the first priority is to understand what parts/products are produced; the next priority is to see if there is a corresponding standard. While this list of part numbers is being assembled, it may be necessary to note, next to each part number, the general category to which it belongs. This notation of product group will facilitate the future generation of summary reports by product group. Utilizing the same cables example, assume that there are four types of cable assemblies, defined as cable types A, B, C, and D. These letters should then be noted next to each part number. These designations should then provide the link between the specific production units and the general units of production or product groups.

The next link that has to be established is that between the manufacturing department's products and the company's end products.

5.4 CONVERTING THE COMPANY'S PRODUCT FORECAST TO A FACTORY PRODUCTION SCHEDULE

Relating what the manufacturing department's produce to what the company eventually ships ranges from exceptionally simple to exceptionally complex. Whatever the case, it is necessary to document the relationship between what the company ships and what the factory produces.

At this point two pieces of product data have been collected. One is the marketing forecast, or at least a list of the company's products in the format that they will be forecasted. The second is the list of items produced by each factory or manufacturing department. A production rate for these factory-produced items must be forecasted to develop a manufacturing plan. Usually a forecast for these factory items must be extrapolated from the marketing forecast. The production control department does this extrapolation for short-range factory scheduling. For short-range scheduling or long-range planning, a ratio of component items per product must be utilized.

Documenting a ratio of component items per product may require either negligible effort or a major study. If a product is static, i.e., it is made with the same mix of components every time, the information may be readily available. However, if the product is dynamic, i.e., the component mix varies with each product shipped, then the process is more difficult. A study may have to be made to determine the component mix that best describes a typical product.

Examples of static products can be seen in process, fabrication, and assembly industries. For example, the component structure does not change for products such as beer, gasoline, or for a particular model of golf ball, ball point pen, or sewing machine. Dynamic products are mostly major assemblies with many options. Examples would include computers, sophisticated production equipment, job shop products, and, perhaps, automobiles. Products in batch industries such as plastic and rubber molding or chemical formulation might also be considered dynamic.

5.4.1 Static Products

Static products can be found in fabrication, process, and simple assembly industries. For these products, the conversion from the market forecast level to the factory's component level is a straightforward procedure. It can be done through product designs, bills of material, and production control systems.

For each end product, a list of its components and corresponding quantity must be constructed. This components list should utilize the factory part number designation. In a factory

Converting the Company's Product Forecast 77

where the forecasted end products share some of the same components, a matrix should be established. Utilizing a standard spread sheet, a list of the component part numbers should be made down the left-hand column. Across the top, each column should be labeled for each forecasted product. Having labeled a column for each product and a row for each component, the intersecting boxes must be labeled to quantify their relationship. For static products, this matrix can be filled in directly from a bill of materials. Usually a bill of materials for these products will be readily available.

The most likely place to start is with the production control department. Since it is responsible for scheduling production to meet shipment demands, its bill of materials will probably be defined in factory terminology. Manufacturing engineering and/or product engineering also should have the same information available. The planner can and should be specific as to what is being requested. Presenting a copy of the blank product/component matrix will clarify and focus the data collection effort. The matrix will show in specific terms what information is needed. Unfortunately for the planner, it is possible that the production control or engineering department will not provide a bill of materials exactly in the format desired. The planner may have to extract the information desired from the existing documentation. The planner may be directed to an official bill of materials, a production controller's notebook, engineering/product drawings, or a computer-maintained data base. Some interpretation or reformatting may be necessary.

When evaluating any bill of materials it is important to understand how it is constructed. For example, there is a big difference between an indented parts list and a summarized parts list. Obviously, it is important that the planner be aware of the format being utilized. An indented parts list describes the composition of each assembly and subassembly. A common part, such as a screw, could show up many times, under different subassemblies within the bill of materials for one product. However, a summarized parts list will only list each part once, along with the total quantity of that part per product. The screw, as represented by a part number, will only appear as a component of the end product. In a summarized parts list, there is no breakdown or description of subassemblies.

Another typical problem is when a long-range forecast is generalized by product type. Within each product type there may be a variety of product models. Most likely any bill of materials will document the configuration of the product models, but not the general product type. It then may become necessary for the planner to develop a ratio of the product models to the product type. In a relatively stable and simple product environment, product model to product type ratios may be easily established. If unavailable from the marketing department, these ratios could also be constructed from shipments history. If developing these ratios is not a simple process, there could be two problems: (1) the product types as defined in the marketing forecast are too broad. Whether from the ratio technique or from the marketing organization, the net effect may be a need for a more detailed forecast by product model; and (2) the products are not really static. If there is fluctuation or a constantly changing relationship between the product models and the product type being forecasted, these product types may be considered dynamic. They are dynamic in the sense that developing a bill of materials is no longer a simple process. More sophisticated techniques must be used to develop a bill of materials suitable for planning purposes.

5.4.2 Dynamic Products

Establishing a bill of materials for dynamic products is sometimes difficult, but for manufacturing planning a specific bill of materials is required for each end product being forecasted.

Dynamic products, by definition, have a missing link between the products as forecasted and the components as manufactured. This gap is, perhaps, most prevalent in industries that produce expensive and complex assemblies. This situation may also occur in a job shop manufacturing environment or where the product has many customer-sensitive options. An example of a dynamic product can be seen with the manufacturer of office trailers, like those used on construction sites. A 5-year marketing forecast may specify the quantity of each trailer type to be produced. It may not specify the mix of optional components such as the number and type of doors to be installed, electric and plumbing options, and both the exterior and interior finishing options. Since these component options affect manufacturing time, it is essential that a link be established between these component options and the product forecast.

Quantifying options/configurations is even trickier with more complex products such as mini-computers. Assume that a 5-year manufacturing plan must be developed for the circuit

Converting the Company's Product Forecast 79

board fabrication and assembly departments. Assume also that the forecast must be based on the demand for each computer type. To determine circuit board manufacturing requirements, it will be necessary to first develop a circuit board forecast. The only way to convert a mini-computer product forecast to a circuit board forecast is to establish a ratio of circuit boards to each mini-computer product as forecasted. Developing these ratios cannot be done simply from a bill of materials. Mini-computer systems are shipped with a wide variety and quantity of options. Two shipments of the same mini-computer product could have a radically different board mix. There may be a wide range between the stripped down version and the fully loaded version of the same mini-computer product. Its mix may be determined solely by what the customer orders.

Whether one's product are dynamic or static, a bill of materials must be established and documented. Having established both the forecasted product groupings and the specific factory component items, a product/component matrix can be constructed. This matrix will look the same as what would be used in a static product environment. The only difference is that when completed, the static product matrix will contain the ratios in whole numbers and the dynamic product matrix will be in decimals. Constructing this matrix, although blank, will illustrate the specific task to be achieved. The matrix can be useful as a focal point for the planner's efforts and as a means to communicate what is needed to others.

In order to fill in this matrix, a number of approaches can be taken, although for any approach the planner must define what additional information is needed. For the planner to be specific, an understanding of the missing link must be established. For example, there may be a clear-cut relationship between the components in question and the subassembly level. However, the subassembly level, whether a product model or option, may not have a direct link to the product being forecasted. Knowing that the subassembly is optional or variable, the planner can then focus on the relationship between the subassembly and the product rather than the component and the product. Simplifying what information is needed should minimize the time needed to collect it.

When marketing simply forecasts the generic product while omitting its makeup, steps must be taken to capture this detail. Essentially, predicting product configuration is an exercise in forecasting. Like any forecasting there is not any one correct way to do it. Any number of approaches can be taken, depending on the type of industry, the company's management's style, or, perhaps, the planner's prerogative. For manufacturing planning, three approaches can be taken to develop this detail:

1. Request additional detail from marketing. In some cases, marketing may be able to provide the additional detail needed without a major effort. Although a top level forecast was published, it may have been constructed from the secondary product model level. Marketing may have rolled up the forecast to the top level to simplify presentation to top management or the financial department. In this case, the level of detail required by the planner might be readily available. It usually does not hurt to ask. However, the forecaster should not be coerced into working with product groupings with which he or she is unfamiliar. In this case, the planner might be better off developing the detail himself.
2. Utilize tracking models. A tracking model may be developed for each product by financial analysts as a fictitious representation of what is a typical version of that product. Tracking models are fabricated in order to facilitate the monitoring of manufacturing costs. Throughout the year, the cost to manufacture each tracking model is developed. This enables the financial group to monitor the effects of manufacturing process improvements, engineering change orders, or production problems. These tracking models may be utilized for manufacturing planning provided they are, in fact, representative of a typical product. However, they usually are not typical representations. The primary purpose of a tracking model is to establish a fixed representation of the product. It is of secondary concern that that model be a typical representation. Unless the products are relatively simple, tracking models should be thoroughly evaluated before being used by the planner.
3. Develop historic data and trends. By reviewing what has happened in the past with each product, one can clarify a picture of that product for planning. When historic data are utilized, two approaches can be taken. One utilizes the current option/model mix when component requirements are developed. The other approach,

Converting the Company's Product Forecast

which may or may not be necessary, is more sophisticated. This is to utilize historical data and project trends in option/model mix for each product. In either case, it is first necessary to uncover the historical data. This information might have to be constructed from raw data, such as the order form, the factory routing sheet, the shipping forms, and the invoice used for each customer's order. In other environments, the information may be available in a more succinct, summarized, and usable format. For example, at the end of each week, production management reports the total quantity of each product model or option produced. Also, these weekly reports may be available for any number of previous weeks, months, or even years. By evaluating the configurations of previous shipments, a typical configuration can be quantified for planning purposes. In most cases, simply determining an average configuration should be sufficient, although for some key items it may be necessary to project trends. Determining which route to pursue should be based solely on the impact on manufacturing. If manufacturing is highly sensitive to product configuration, then the configuration mix to be used for planning should be based on a thorough analysis. If, however, the product's configuration is relatively stable or does not significantly impact manufacturing, then a "ballpark" configuration mix may be all that is necessary.

Table 5.1 is an example of a components per product matrix for a group of dynamic products. The objective is clear: to establish the quantity of each component per each product. Unfortunately, it may be difficult to document how many #004s there are in product B. Especially if 004s are optional or part of an optional subassembly, then filling in the matrix may become a project.

Assume that the products A, B, C, and D are four major types/sizes of office trailers. Also assume that part #004 designates a window. Within each product type, there are models ranging from stripped to deluxe. Deluxe does not necessarily mean more windows; it may mean a bathroom, therefore less windows. For instance, product A comes in three models: economy, special, and deluxe. Economy comes with only three windows.

Table 5.1 Component per Product Matrix

Component part #	Product			
	A	B	C	D
#001	1.00	1.00	1.00	2.00
#002	1.00	1.00	1.00	1.00
#003	0.90	0.90	1.00	1.00
#004	4.10	4.70	2.50	–
#005	–	–	–	1.00
#006	–	0.50	0.50	0.50

Special has six windows and deluxe has four windows and a bathroom. To determine the number of windows per office trailer type A, one has to determine the percentage of sales by product model. In this case, 50% of the A trailers sold are sold as economy models. The special and deluxe models sell at 30 and 20%, respectively. Therefore, for every A trailer that is forecasted, the planner should expect 4.1 windows to be installed (Table 5.2). The ratio of 4.1 004's per product A can then be entered on the component per product matrix. In this example the product models are static, i.e., the component mix is fixed and documentation is readily available. Therefore, the only dynamic area is the relationship between the product models themselves and the overall product group as forecasted. Once this relationship is established, the component mix for the forecasted product can be calculated easily.

Table 5.2 Calculating Component per Product Ratios

Product model A	Ratio of sales		Windows each		
Economy	0.50	×	3.00	=	1.50
Special	0.30	×	6.00	=	1.80
Deluxe	0.20	×	4.00	=	0.80
Average number of windows/product				=	4.10

Converting the Company's Product Forecast 83

In the world of mini-computers, the product variations are exceptionally complex. There might be model numbers within and across product groups used to designate the major optioned and fixed items such as the expansion capability, or communications options. Manually calculating and developing a components per product matrix may be a major undertaking. However, through the use of computers the task may become manageable.

5.4.3 An Automated Product Configuration System

In a rapidly changing complex product environment, automating the generation and maintenance of a current ratio of components per product may be beneficial. Such environments are usually the producers of highly sophisticated products. These products consist of many optional components. Also, because of rapidly changing technology and customer demands, documenting a "typical" product configuration is difficult. This is the situation for a major computer manufacturing business. For planning, it has become important to utilize up-to-date ratios of components per product.

Determining the components forecast from a mini-computer product forecast (12 mini-computer product types) is not an easy process. These components are not related directly or simply to the company's end products. The products are shipped with a wide variety and quantity of options as ordered by the customer. A components forecast must incorporate the option configurations desired and add-ons or expansions to previously shipped systems, as well as product forecast.

The first concern is to develop a components per system ratio that is a reflection of a current "typical" system. A medium- to long-range components forecast is to be based on educated expectations of systems, options, expansion configurations, and product demand. For the long range, customer orders are unavailable to construct the circuit board, cables, and subassembly forecasts.

To develop and maintain an up-to-date, accurate ratio of components per system, it is necessary to look at past shipments. Other methods for determining components per system were deemed unsatisfactory. Configuration ratios based on future bookings, could be skewed by a major customer ordering many systems of the same configuration. Financial tracking models are static for an entire year and are not necessarily a reflection of a typical system. A typical system can be developed from shipment history where, during a given time frame, there is a mix of systems for both minor and major customers. Unfortunately, no records are kept of the components that have been shipped within each system or expansion. There is, however, a monthly billing report that details each shipment's configuration with higher level identifiers. These identifiers describe a subassembly level that may be made up of one or two circuit boards, subassemblies, and cables, and are known as internal product identifiers (IPI). IPI can be accurately broken down to the component part number level through equivalency tables, bill of materials, and product drawings. These IPI are then bundled under the marketing identifier (MI) level. The monthly billing report contains a record of each customer order, and its configuration to the IPI level. For example, a typical customer order will contain a few MIs, which will designate major units such as the central processor, and major options such as printers, tape drive, and additional memory. Underneath the MI are the IPI which are called out by that MI. Some of the IPI are conditional or optional; they do not necessarily ship with the MI every time that the MI is shipped. Utilizing actual customer orders, rather than a bill of materials, will allow these conditional IPI to be quantified by history.

Table 5.3 is an example of the data contained and utilized from the billing detail report. In this case, the product shipped was a DPS 6/32 mini-computer. The MIs designate a central processor (CPX9632), a printer adapter (PRM 9630), and additional memory (CMM 9630). The 12-character IPI designate a subassembly level such as

BCPU137A-002 = Central Processor Circuit

BCMM0448-001 = Memory Circuit

BPWU600A-001 = 60 Hz Power Supply

The same data base that generates the monthly billing report is used to create a ratio of IPI per product (mini-computer systems). Each month data are summarized and stored. Individual customer orders are of no concern, so only the total shipments of each IPI by product and the total shipments of each product are tabulated and stored. This new data base is

Table 5.3 Billing Detail Report Data (Edited)

Customer order #NB 265650
Product = DPS 6/32
	Quantity

CPX 9632	
BPNL150A	1
03910313-005	1
BSMF004B-002	1
60134360-004	1
60143545-001	1
BCPU137A-002	1
60143490-003	1
BDSK002A-001	1
BCPUD37A-003	1
BCAB005C-002	1
60134444-022	1
BBDC001B-002	1
BMMU031B-002	1
BMLC011B-002	1
60143370-001	1
60143368-001	1
BSMF002E-001	1
BRCK023A-001	1
BCMM044B-001	2
BPWU600A-001	1
60143516-003	1
60129892-020	1
BMLFCLAA-001	1
60135450-001	1
PRM 9630	1
BDCEPRTB-002	1
60143586-001	1
CMM 9630	1
BCMM044B-001	2

capable of storing, by month, 36 months of IPI and product shipments data. It should be noted that the IPI shipped as expansions or add-ons to previously shipped systems are also collected and stored by product.

The manufacturing planning system utilizes a 5-month rolling average of IPI per product. This file is recreated each month along with the data base itself. As each month ends, it and the previous 4-month shipments of IPI by product are summed. The 5-month total shipped of each IPI to a product is then divided by the total shipments of that product. The IPI must relate to the product as defined in the marketing forecast; in this case, the product forecasted is a mini-computer system. Any customer order that contains a complete system, as defined by the presence of a central processor, is tabulated to develop the count of product shipments.

New product ratios are entered manually. Obviously, new product configurations must be entered manually since there have been no shipments.

5.4.4 New Products

Some special techniques may be necessary to successfully generate a useful bill of materials and incorporate a proposed new product into the manufacturing plan. There are two approaches that can be taken. One is to generate the manufacturing requirements for a new product separately and then add them to the manufacturing plan. The second approach is to treat the new product like an existing product and integrate it into the planning process.

The approach taken will depend on the new product itself. A new product that is a radical departure from the existing product line will, of course, require separate study. If the new product is similar to existing products and will utilize the same manufacturing techniques and facilities, it can be handled much like an existing product.

For a new product, it is necessary to construct a bill of materials. The information will most likely come from the product design engineers, although in cases where the product horizon extends beyond 5 years, it may be necessary to consult the marketing department. In this case, either a bill of materials must be estimated or an existing product used as a representation. Like any product, the bill of materials for a new

Product Design Changes and New Technologies

product must be representative of a typical configuration. The need for configuration detail must be emphasized when the component mix of a new product is being established. Although the design and marketing functions are interested in product manufacturing costs, there is less concern with the product's overall impact on manufacturing. They must understand that the manufacturing plan and the business depend on a well-conceived projection of a typical bill of materials.

Once a bill of materials has been established, it must be evaluated. If the new product's components are currently understood or produced, then the new product has been adequately defined. If, however, the new product has new components, then additional refinement is necessary. In many cases when new components are similar to existing items, the existing item should be used as a substitute. When an existing component cannot be substituted to represent a new component, the impact of that new component on manufacturing must be studied separatetely.

For most manufacturers, an accurate manufacturing plan can be done by utilizing existing products or components as substitutes for new products and components. From a manufacturing standpoint, the new product(s) could be quite similar to existing products. Many times a manufacturer will introduce a new product in which both the design and the materials are new. A golf club manufacturer may introduce a new line of clubs with a radically redesigned grip, a new metal alloy shaft, and newly designed heads. While the new golf clubs may be a significant departure from the norm with regard to design, materials, and marketing, its assembly process may remain the same. For the purpose of assembly-line planning, the use of the substitute technique would, in fact, be quite acceptable.

5.5 PRODUCT DESIGN CHANGES AND NEW TECHNOLOGIES

Marketing may forecast the current product mix into the future. While the product line may remain the same on the exterior, the planner must concern him- or herself with important changes to its interior. Although subtle externally, significant interal alterations may be made on the product over time. Design changes are usually made to reduce material or manufacturing costs or increase quality and reliability. Many times these changes will utilize new technologies.

The planner must consider the product design changes only with respect to their impact on manufacturing requirements. If the product design changes and new technologies are essentially conceptual, their inclusion in the manufacturing plan may be risky. It is better that the plan be based on current product design, which is a proven rather than a hypothetical concept. In the other extreme, if the product design changes are a "sure bet," they probably should be included. For manufacturing planning, the definition of the product should be based on the best substantiated information available.

The prediction of product design changes, in many cases, involves technological forecasting. If this forecasting is done by marketing or product design engineers, perhaps it should be incorporated into the planning process. To utilize the forecast, it must be documented in sufficient detail. Detail is necessary if the planner is to accurately adjust a product's bill of materials to reflect the design change. A specific example can be seen in the computer industry. A computer's circuit board requirements would reflect the use of 64K memory chips. It is projected that within the next 2 years they will be replaced to a great extent by 256K chips. To be specific, the introduction of 256K chips would mean a reduction of memory boards from three to one in 2 years. (A complete replacement of 64K chips might take 3 years.) To reflect this technological impact, the computer's bill of materials must be altered. An accurate representation could be done by reducing the computer's memory board requirements to one third of what is currently needed. This approach assumes that both the 64K- and 256K-based memory boards utilize essentially the same manufacturing requirements. The 64K board would then act as a substitute for the 256K board under development and the reduced quantity would reflect the impact of fewer boards required.

Actually there are two ways that the impact of changes in component structure can be handled: (1) to factor the ratio of components per system, and (2) to wait until a component forecast is developed and then factor that forecast. The planner may want to consider the latter approach if the changes are to be introduced gradually over a period of time. Take the example of the 256K memory boards replacing the 64K memory boards. Perhaps in the first year, only one quarter of the 64K memory board requirements will be replaced by 256K boards. It may be 4 years before all the 64K boards are phased out.

Projected product design changes and new technologies do require scrutiny by the manufacturing planner. If incorporated into the manufacturing plan, it is imperative that these assumptions be both substantiated and well documented. Before their inclusion, one may want to review these assumptions with the people who have requested the plan. This is important since it is typical that a quantified technological forecast will be controversial.

There is one other alternative that should be considered. This is to develop a manufacturing plan that is based on current product design and another based on proposed design. This two-scenario approach may or may not be a realistic alternative, depending on the complexity of the change and the complexity of both the product and the process. However, if this approach is feasible, the planner can remove him- or herself from the controversy and simply present the facts. The decision making then will lie where it belongs—with upper management.

Usually the manufacturing planner will come from a process-oriented background. That is, he or she will be an industrial engineer or a manufacturing engineer. The planner should deliberately avoid making assumptions or decisions regarding the product to be used for manufacturing planning. The emphasis should be placed on collecting information from others who are responsible for product design engineering and planning.

REFERENCE

1. I. Hill, Avoiding pitfalls in long-range facilities planning, *Proceedings of the Spring Conference of the American Institute of Industrial Engineers*, 1981.

6
Manufacturing Process Data

6.1 INTRODUCTION

Although the collection of product data may be outside the planner's expertise, the collection of data on the manufacturing process should place the planner in familiar territory. This familiarity will enable the planner to collect and utilize the manufacturing process data more efficiently. The manufacturing process is essentially the interface between the materials, the men, and the machines. This interface must be defined and documented. The first step is to accurately definine the manufacturing operations and their sequence. After each step of the process has been documented, the relationship of manpower and equipment to each of those steps can then be established.

6.2 MANUFACTURING OPERATIONS AND SEQUENCE

An accurate manufacturing plan must account for every step of the manufacturing process. It is also important that the sequence of operations be established in detail. Although the establishment of such a sequence may seem obvious and simple, in some cases it is not. Only through a detailed examination of the manufacturing process will many of its subtleties and previously hidden idiosyncrasies become apparent. The objective is to ensure that the manufacturing plan is based on the operation as it really functions.

Documentation of this level of detail may also generate some controversy. Controversial revelations during this process are typical and may actually be beneficial. It is better that any disparity between what is actual and what is perceived of the manufacturing operation be cleared up at this early stage. The alternative is to have controversy at the final presentation of the completed manufacturing plan.

The operations and their sequence are the foundation of the manufacturing plan. The rest of the planning process consists of adding building blocks to that foundation. It is imperative that this foundation be both structurally sound and accepted.

6.2.1 The Process Flow Chart

The process flow chart is the best method for documenting manufacturing operations and their sequence. If it is done properly it will be accuarte, detailed, and very presentable.

A flow chart should be developed for each major manufacturing area. It should be done in detail, with each step of the process clearly defined. A detailed flow chart should include not only the operations themselves, but every transport between operations, every predictable delay (drying or curing), inspections, and storage. These steps should be noted on the chart, utilizing the standard process flow chart symbols illustrated in Table 6.1. Although each step of the process is briefly labeled, they should also be numbered. A numbering scheme will enable backup sheets for each operation to be easily cross referenced with the flow chart. Although not standardized, it is sometimes helpful to designate each step by its type of activity. For example, the operations could simply be numbered and the transportation steps designated by the letter T and then a number. Similarly, if there were 12 delay steps, they would be labeled from D1 through D12. An example of a process flow chart can be seen in Table 6.2.

While a detailed flow chart is being developed, some accompanying data should also be collected. The flow chart

Table 6.1 Process Chart Activities[a]

ASME symbol	Name of activity	Definition of activity
○	Operation	An operation occurs when an object is intentially changed in any of its physical or chemical characteristics, is assembled or disassembled from another object, or is arranged or prepared for another operation, transportation, inspection, or storage; an operation also occurs when information is given or received or when planning or calculating takes place
□	Inspection	An inspection occurs when an object is examined for identification or is verified for quality or quantity in any of its characteristics
⇧	Transportation	A transportation occurs when an object is moved from one place to another, except when such movements are a part of the operation or are caused by the operator at the work station during an operation or an inspection
D	Delay	A delay occurs to an object when conditons (except those which intentionally change the physical or chemical characteristics of the object) do not permit or require immediate performance of the next planned step
▷	Storage	A storage occurs when an object is kept and protected against unauthorized removal

[a]As standardized by the American Society of Mechanical Engineers (ASME).

Table 6.2   Process Flow Chart

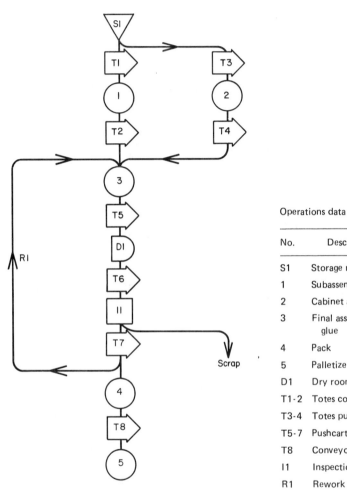

Operations data

No.	Description	Quantity
S1	Storage rack	5
1	Subassembly	4
2	Cabinet assembly	1
3	Final assembly and glue	2
4	Pack	2
5	Palletize	1
D1	Dry room	1
T1-2	Totes conveyor	1 conv.
T3-4	Totes push cart	80 totes
T5-7	Pushcart	45 carts
T8	Conveyor	1
I1	Inspection	2
R1	Rework	—

Manufacturing Operations and Sequence

should present a realistic picture of the manufacturing process. In doing so, rework loops and scrap loss points should be included. the percentage of product which is reworked or scrapped could also be noted on the chart.

The process flow chart will typically have offshoots that deviate from a straightline flow. These offshoots may illustrate the additional manufacturing steps required for certain product models or component types. At some point it will be necessary to determine the volume of production which passes through that offshoot.

Depending on the industry, it may also be essential to note production lead times. The time it takes from when the components are introduced into manufacturing until the manufactured unit is complete should be recorded. Usually a detailed study is not necessary; the accepted rule of thumb used by production management and production control should be sufficient. This information will be needed later in order to accurately forecast manufacturing requirements. For example, if the throughput takes 13 weeks, then the manufacturing requirements must be in place and production started one fiscal quarter before the final product's shipment date.

The primary contributor to lead time is any delay in material processing. In most industries, the actual processing time is significantly shorter than the throughput time. Some delays are an expected and necessary part of the manufacturing process. Such delays, indicated by the "D" symbol in the flow chart, are usually required for curing, cooling, drying, etc. The time required of each product at each delay step should be recorded.

Other delays are due to a lack of manufacturing control. Work-in-process material may sit idle between operations for no apparent reason. If there is storage space specifically set aside for work-in-process, then it should be designated on the flow chart. A storage symbol should be used—not a delay symbol—since there is no specified delay time.

During the development of the process flow chart, it is not unusual for transportation steps to be troublesome. Perhaps this is because there are so many of them that they are taken for granted and are somewhat inobtrusive. Nevertheless, the transport method should be designated and labeled on the flow chart. Additional transport data should also be collected but documented elsewhere. The additional data could include transport capacities, i.e., units per tote box, conveyor capacity, units per pallet load, etc.

The major manufacturing steps, i.e., operations and inspections, require further study. The collection of data on these steps will be covered in more detail later in the chapter.

Once the first rough draft of the flow chart is completed, it should be reviewed by manufacturing engineering and production management. It may also be helpful to have the flow chart reviewed by the line supervisors. They are the ones most likely to recognize the small, nonetheless significant, errors in the flow chart. Also, their perception of the manufacturing process is most likely based on the way things really are rather than on the way things should be. Once the flow charts are reviewed and corrected, they should be circulated again before the final print is drafted. All too often, previously unnoticed errors will be detected. In any case, the planner should be confident that the flow chart is detailed and accurate upon its completion.

6.2.2 Backup Data Sheets

Once the flow charts are completed, specific and detailed process data must be gathered. The objectives of the manufacturing plan will determine what data are necessary. With these objectives in mind, the planner must give consideration to the collection and organization of the pertinent data. To facilitate the data collection effort a standardized backup data sheet should be constructed such as the one illustrated in Fig. 6.1.

A data sheet should be tailored to the objectives of the manufacturing plan. For example, the data sheet illustrated in Fig. 6.1 was designed for a specific manufacturing plan. The primary objective of that plan was to support a 5-year facilities rearrangement plan. Hence, the data sheet emphasizes space and utility requirements, although it should be noted that in order to get the space requirements, information is also needed on labor standards and processing times. However, had the plan's emphasis been financial, the data sheet would have been constructed quite differently. A sketch of the equipment may not have been necessary nor the detail on utility requirements. Emphasis would have been placed on purchase cost, installation costs, maintenance costs, etc., although, like the facilities orientation, the financial-oriented plan depends on data common to all manufacturing plans. Specifically, all plans need similar data for the development of the labor and equipment requirements.

Manufacturing Operations and Sequence

DEPARTMENT _____ OPERATION _____

FOOTPRINT AND TEMPLATE:

SCALE: 1/4" = 1" OR 1/8" = 1"

PRODUCTION DATA

PRODUCT	OPERATION # AND DESCRIPTION	GRADE	STD. TIME	OUTPUT/SHIFT

EQUIP. REQUIREMENTS: HEIGHT: WEIGHT:

WATER _____
STEAM _____
DRAINS _____
EXHAUST/VENTILATION _____
SUPPORTS OR FOUNDATIONS _____
SECURITY _____
ENVIRONMENTAL CONTROLS _____
POLLUTION CONTROLS _____
MAT'L HANDLING EQUIP. _____
OPERATING CONTROLS _____
ELECTRIC _____
AIR/VACUUM _____
SAFETY _____
OTHER _____

PRODUCTIVITY FACTORS

PRODUCT	YIELD	UP TIME	PERF.	UTIL.

AQUISITION FACTORS

PURCHASE COST: _____
INSTALLATION COST: _____
LEAD TIME: _____
CONTACT: _____

NOTES _____

QC 7214

Fig. 6.1   Standardized work station data sheet.

Once a backup data sheet has been formatted, it can be filled out for each step of the manufacturing process. Having each step defined and numbered in the flow chart and its corresponding data sheet will enable easy cross reference. Developing a formatted backup data sheet serves two purposes: (1) it facilitates the organization and management of the large volume of data required for manufacturing planning, and (2) perhaps more importantly, it serves to clarify one's direction. A well-conceived data sheet will focus the collection effort on only the data that are essential.

6.2.3 Plant Layouts

Although not a substitute for a flow chart, a quarter-inch scale plant layout can be useful for the collection of much pertinent information. In fact, it may even be a useful tool for developing the flow chart. In many cases, it is easier for people to relate to a "picture" of the factory than it is to relate to a conceptual diagram such as a flow chart. The planner, when working with others, may find it easier to highlight the flow of materials from a plant layout. Transposing the layout flow patterns into a process flow chart can always be done once the information is established.

Plant layouts, preferably quarter-inch scale, will also enable the rapid collection of data throughout the planning process. A quarter-inch scale layout is especially helpful when the plan's objective is oriented toward space or facilities requirements. This application will be discussed in detail in Chapter 7.

6.2.4 Productivity Improvement

Although not directly related to the development of a manufacturing plan, productivity improvement ideas should be recognized and noted. A manufacturing plan requires intensive study of the manufacturing process. During this study many areas with productivity improvement potential may surface. These areas should be noted and suggested for further study.

At this stage, a detailed overview of the manufacturing operations and their sequence has been documented. This overview may highlight inefficient process flows such as hidden loops

Labor Planning Data 99

and bottlenecks. It may clarify the need for new methods or the ideal application of new technologies. At this early stage of the planning process, the transport steps may demonstrate the biggest potential for improvement. When there are a lot of transport steps, there is usually room for improvement, especially since material handling is expensive. Through detailing the manufacturing sequence and the steps through which the product flows, some new ideas for handling may be obvious. These ideas should be recorded.

Improvement potential may become apparent throughout the planning process. It is easy to get caught up in the potential of some of these improvements. While the potential may be great, the planner should avoid becoming sidetracked. However, the improvement ideas should not be lost. They should be noted, passed on to others, and integrated as a goal into the manufacturing plan.

6.3 LABOR PLANNING DATA

Developing direct manpower requirements depends on knowing what tasks are needed and the time it takes to manufacture each product or its components. The most likely source of this information is from the industrial engineering labor standards data. In the event that engineering standards are unavailable, alternative action can be taken to generate a substitute.

In addition to standard data, labor planning must include a profile of the work force. A labor profile could range from a synopsis of the current head count to the detail provided by a sophisticated labor reporting system.

6.3.1 Industrial Engineering Labor Standards

During this data collection phase, access to and comprehension of standards data must be established. Industrial engineering standards data quantifies the relationship between the product and the process and is, therefore, the major ingredient of a detailed manufacturing plan.

Each operation, as defined in the process flow chart, must be studied. The standard labor processing time for each product or component that is processed through that operation must be recorded. The complexity of the products and the process will determine the most feasible means for organizing this information. Before it can be organized, the standard data must be understood. There are some typical problem areas to be aware of. Problems occur when there is a discrepancy between the process definition of an operation and the labor definition of an operation. In the process flow chart, an operation is a processing step where there is value added to the product. However, the labor standard may define an operation solely by the tasks of the worker. A good example of this can be seen in the rubber or plastic molding industries. A final molding operation will be designated as such in a process flow chart. However, the operation may use a person to load the molds and another person to unload the molds. Hence, for labor measurement, these two operations, loading and unloading, may be distinctly separate. They may even be on separate pay scales, and may even require more people to load than to unload or vice versa. Care must be taken by the planner to note these problem areas for future reference.

A similar concern arises when one worker handles more than one manufacturing operation. With the increasing prevalence of robotics and other automated equipment, this situation will become the norm.

Those operations where a standard has been set to handle a given lot size must also be noted. For instance, in a mold with 40 cavities, the accompanying standard data would most likely assume a lot size of 40 units. Lot size may also be a consideration in a standard because of material handling technique, economic processing quantities, or packaging.

Another problem can occur when a planner is solely interested in a bottom line manload projection. The planner might then request the total standard time required of each component or product. This total standard time may or may not include the subassembly labor time. If not, the labor time for each subassembly must also be collected and incorporated. Conversely, the planner must recognize when subassembly labor times are included.

This awareness of the difficulties encountered with labor standard data is necessary to ensure that the information is properly interpreted and utilized. In any case, for detailed planning a standard labor processing time must be collected for each item processed through each operation. These data should be available in detail from the industrial engineering department.

Labor Planning Data

Standards data may also be available from a computerized labor reporting system or the cost accounting department.

In some cases, it may be essential to access some of the backup data which were used to formulate the standard. If equipment requirements are to be generated, or the standards adjusted for specific productivity improvements, then more detail is necessary. One source of this information would be the backup sheets with the elemental times used to develop the labor standard. Utilizing the elements and associated times which make up a standard will be discussed in Section 6.4.

Once the standard data source has been established, the standards must be collected and organized to allow the information to be usable as well as presentable. The planner must ensure that all the pertinent standard data are available and complete. The backup process data sheet, which corresponds to each operation on the flow chart, can be used to organize standard data. Organizing the standard data by operation is applicable when a limited variety of products or components pass through each operation. Each item that is processed should be listed on the data sheet. Corresponding to each item should be the standard time, in hours, that it takes to process that item through the operation.

In cases where the operation has no labor content, a notation should be made on the data sheet to that effect. In cases where the operation contains more than one labor function, two approaches can be taken. One is to sum the two standards in order to quantify the total labor content of the operation. The other is to note each labor function separately on the data sheet. For example, this situation might occur when the loading and the unloading of a machine are independent tasks with independent standards.

A process data backup sheet is not always the best format for recording standard data. In some situations, the standard spread sheet may be a more useful means for organizing standard data. A spread sheet becomes necessary when there is a wide variety of components or products that are processed through the same operations.

The spread sheet's columns may be labeled for each product or component and the rows may be labeled for each operation or vice versa. The format selected depends on one's preference and the number of operations versus the number of components.

The operations noted on the spread sheet should correspond to those illustrated on the flow chart. Also, this spread sheet should contain the number of each operation as it is designated on the flow chart. In cases where the processing operation contains more than one labor operation, the planner may wish to amend the operation number as it is recorded on the spread sheet. For example, a molding operation may be designated as operation 2, yet for labor content 2A might designate loading, while 2 B might designate unloading.

Computer Files

In many companies, industrial engineering standard data are maintained on a computer file. If such a file is to be the primary source of labor standard data, the planner must understand its content, format, and accessibility.

Effective planning requires easy access to reference data such as the standard times. Unfortunately, many computer files, although accessible, are unintelligible. To be usable for the generation of a manufacturing plan, the standard data must be properly organized. If the information contained in the file is complete, reformatting the file may provide the clarity and organization desired. Depending on the file's construction, the generation of a reformatted version may not be difficult. The planner should request a format similar to that of a spread sheet.

Sometimes it might be simpler for the planner to construct a spread sheet manually from the data provided by the computer file. Whether accessed from a computer terminal or from a hard copy printout, the planner must understand the file's format. Then this difficult-to-use data source can be transcribed into a useful reference document.

When a computer-aided planning system is designed, a reference document may not be necessary, although it is still important for the planner to know what information is available on the file and in what format it is available. The planner should check that, for each operation in the process flow chart, any corresponding direct labor time is recorded in the standard file. A computerized planning system does not require the same visually organized data that are required for a manual planning system.

The planner must construct an easily accessible and readily interpreted source of standard data. Standards are essential to the success of a one-shot manufacturing plan or a sophisticated computer-based planning system.

Labor Planning Data

6.3.2 Simulating Standard Data

There may be instances where standard data are unavailable or unusable, yet standard data are essential for manufacturing planning. In this case it will be necessary for the planner to construct the labor standards, although for applicable planning purposes, planner-generated standards will lack the detail and refinement of engineered standards. Engineered standards usually require the full-time efforts of an industrial engineering department to develop and maintain. The planner can, however, rapidly develop a substitute for standards data which may be adequate for planning purposes. These substitute standards can then be entered on the flow chart's data sheets or on a spread sheet the same as an engineered standard. Instead of an engineered standard, a reasonable expectancy time will be substituted. The expected processing time for each component or product through each operation can be calculated from actual performance.

Reasonable expectancy time can be calculated for each operation by dividing the man-hours worked by the production output. The planner must be sure that the expected production output is reasonable and accurate. It should also reflect the operation's peak realistic capacity. If possible, this information should be based on actual experience over an extended time frame.

The best source of this information may come from an analysis of direct labor production reports. In a sophisticated environment, these data may already be summarized on a routine production report. A labor report may clearly note the hours worked and the number of units produced for each operation.

In a less organized environment, the expected daily production rates may have to be obtained from production supervisors or, perhaps, the operators themselves. If this approach is necessary, the planner must make every effort to explain why the data are needed. An atmosphere of cooperation and trust will enhance the value and accuracy of their contribution.

6.3.3 Other Labor Data

Beyond standard data, a profile of the current labor force should be available to the planner. Specifically, the current head count and other labor statistics will be needed. During this data collection phase, the planner should determine what information is available and where it can be accessed.

It is inevitable that the planner will need a breakdown of the current head count. A breakdown of head count would include the staffing levels of each department or work center. It might also include a breakdown of the work force by labor grade. At the least, a distinction should be made between direct and indirect labor.

A profile of the current labor force may be used later as a reference to compare projected head counts. Such a comparison will enable the planner to test the reasonableness of the manufacturing plan. More importantly, the difference between the current head count and the projected head count will quantify the number of employees that should be hired (or laid off). Current head count information may also be used to establish the percentage of indirect labor.

In addition to gathering information on the current head count, the planner should also develop other sources of labor information. The planner should understand labor reports that may be utilized by production management or the financial department. At this point the planner should become aware of the existence of these reports and a general understanding of the data they contain. In general, these reports may contain specific data on the productivity of the work force. These productivity data may be needed later to ensure that future resource requirements are, in fact, based on realistic expectations of the worker.

6.4 EQUIPMENT PLANNING DATA

Collecting equipment planning data is very similar to collecting labor data. The bottom line is knowing how much equipment time is utilized when processing each product or component although, unlike labor standards, equipment standards are usually not readily available.

In most cases, the equipment processing time for each product can be determined from the labor standard. When the worker and the equipment are mutually dependent to produce, the labor standard is the same as the equipment standard. The output per man-hour is the same as the output per equipment-hour.

Equipment Planning Data

In slightly more complicated operations, the equipment-hours can be extrapolated easily from the labor standards. Extrapolations can sometimes be made when more than one worker is required to operate a piece of equipment; one worker can handle two or more machines at once; or when a worker loads a machine and unloads it hours later. For instance, when a machine requires two full-time operators, the combined labor standard divided by two should yield an equipment standard. Specifically, if a machine processes 100 units per hour, the equipment standard is 0.01 hours per unit. However, if one operator loads the machine (at 100 units per hour) and another operator unloads, their combined man-hours equal 0.02 hours per unit. Therefore, the equipment content per unit is one-half the labor content.

Another source of equipment standard data would be from the manufacturing engineering department. Typically, manufacturing engineering is involved with the mechanical aspects of the production process. Manufacturing engineers are usually involved with both the processing and the handling equipment. Their equipment orientation should enable them to produce information on equipment capacity. Capacity data might come from the equipment's specifications, production studies, or trial runs.

Although helpful, the planner should be careful of capacity specifications. There may be a significant difference between a piece of equipment's theoretical capacity and its actual capacity. In many cases, equipment capacity is constrained by the operators, the items being processed, the plant layout, job design, or the manufacturing method. The planner must recognize and adjust for any situation-specific idiosyncrasies that could alter the capacity of the equipment.

Like labor standards, the equipment standard should be an engineered figure. It should reflect the optimal and normal processing rate of the equipment. For example, load and unload time is one aspect of the normal operation of the equipment and should be included in any standard. However, time spent idly waiting for materials or downtime may not be considered in the equipment's production standard. It should reflect a steady, efficient, but realistic working pace.

For each operation, the equipment standard data should be handled much the same as labor standard data. They need to be organized in a usable and presentable fashion.

If the variety of units processed is sufficiently small, the equipment standard data may be recorded on the operations data sheet described earlier. Utilizing these data sheets will ensure that each operation is evaluated.

If the variety of units being processed is large, the spread sheet approach may be necessary. The spread sheet format should be the same as that used for labor standards. The operations and products should be recorded on a matrix. An equipment standard should then be noted for each product at each operation.

When a computer-aided planning system is being developed, equipment data may be difficult to access. Chances are that they will be unavailable on computer files. However, a computer system can extrapolate the equipment standards from the labor standards the same as would be done manually. Extrapolation is possible when there is a straightforward relationship between the operators and the machines. Many operations by design will require a set number of machines per operator or a set number of operators per machine. This situation will enable the equipment requirements to be calculated from the labor requirements. Equipment which cannot be mathematically related to labor will require that its production rates be independently maintained by the computer system. Computer-aided planning will be discussed in more detail in Chapter 14.

In addition to collecting the processing time for production equipment, it may also be necessary for handling equipment. As a vital link in the production process, handling equipment has become more sophisticated and more necessary. Primarily, the emphasis on handling equipment is due to increasing labor costs, material costs, and the competitive market place. Since handling adds no value to the product, it must be done the cheapest way possible. This concern for minimizing costs, combined with new technology, has led to an increasing dependence on automated handling equipment, ranging from "smart" conveyors to robotics.

A dependence on handling equipment makes it imperative that major handling equipment be included in the manufacturing plan. Generating handling equipment requirements will necessitate calculations based on the equipments' handling capacities. However, in many cases the handling capacity may far exceed the production equipment's operating capacity. Handling equipment requirements will then be based on physical limitations and plant layout rather than on considerations of its capacity.

For handling equipment, the notation of any revelent data may also be done on an operations data sheet. The information can then be easily referenced later in the planning process. A

Equipment Planning Data

separate data sheet for major handling operations will also ensure that highly automated and expensive handling equipment is not overlooked.

Not only must production and handling equipment capacities be recorded, but a multitude of other pertinent information should be considered. Where relevant to the objectives of the manufacturing plan, an equipment profile should be recorded on the operations data sheet. For each operation, a record of the number of pieces of equipment available should be recorded. Later, a comparison of the current assets and future needs will highlight the additional equipment required. It will also help in the evaluation of alternative manufacturing strategies, which will be covered in Chapter 13.

A record of each piece of equipment's age and its expected life may be helpful. Also, equipment cost may be needed, especially if the plan's orientation is financial. The incorporation of an equipment replacement schedule into an acquisitions plan could increase significantly the funding required.

The timing of any funding is also dependent on the equipment's lead time. Some highly specialized and sophisticated production equipment may require up to a year between the time it is ordered until it is installed and operational. This lead time should be documented.

Another consideration not to be overlooked is the support equipment required by each operation. The need for major support equipment should also be noted on the operations data sheet. Support equipment is particularly important when the equipment is expensive or takes up space, such as pollution control equipment, generators, and compressors.

If the emphasis of the manufacturing plan is on facilities and future plant layout, then utility requirements should be defined. A notation could be made on the operation's data sheet detailing such items as the electric service required, and other services such as air, water, steam, drains, or any other special requirements.

The amount of information that can be collected on a piece of equipment appears to be endless. The planner must make an effort to keep the objective of the overall plan in mind when developing these data. Maintaining this perspective will minimize any effort in obtaining data that are essentially irrelevant. The planner must determine what information is vital to support the plan's goals.

Data on equipment, labor, and the overall manufacturing operation are needed to construct a manufacturing plan. In order to be useful, the data must be complete, but not excessive. They must also be organized; therefore, easily referenced. Stated simply, the planner must define what data are needed, define what data are available, collect or generate the required data, and organize those data.

7
Facility Data

7.1 INTRODUCTION

Facilities data may be defined as that information which is necessary for projecting square footage requirements. Collecting this information is essential if the manufacturing plan is to be accompanied by a plant layout or even a gross projection of the total space required. The level of detail needed depends entirely on the plan's overall objectives. A manufacturing plan done to support a financial plan may require only a summarized projection of the total space required. However, if the plan is done to support factory reorganization and layout considerations, a detailed quarter-inch scale layout may be necessary. While base data are the same for both requirements, the latter will require information that is more specific and detailed.

7.2 EQUIPMENT SQUARE FOOTAGE

Production equipment is the primary determinant of a factory's square footage. Any projection of floor space requires a knowledge of the number of pieces of equipment and the space taken up by each piece. At this stage, the data on the square footage per piece of equipment must be collected and organized. Each machine or piece of equipment occupies space. In addition, each machine or piece of equipment demands a minimum work area in order to be functional. It is this area that is most pertinent to the manufacturing plan.

The criteria used to define a minimum work area for each piece of equipment must be consistent. It must also be realistic. Equipment space can be defined as the minimum floor space required for the equipment/operation to run comfortably and efficiently. This area or "footprint" as seen in Fig. 7.1 should include the space needed for the operator, work-in-process, material handling devices, minor aisle space, supplies, maintenance areas, and support equipment:

1. The operator. The operator of the equipment takes up space. A footprint must include the operator space that is essential to the operation of the equipment. The operation of the equipment depends on access to controls, materials, and means of egress, therefore, it becomes a part of that equipment's space requirements.
2. Work-in-process. By definition, an operation or inspection station processes material. Material may take up space within the work area both before and after it is processed. The "footprint" must reflect adequate space for this work-in-process.
3. Materials handling devices. Handling devices, such as a conveyor, may be an integral part of the operation. The portion of the handling device that is needed to service the operation and its immediate work area should be included. For example, there are many situations where a conveyor is shared by work station on either side. (See Fig. 7.2.) Therefore, only one-half the width of the conveyor need be included in the footprint of a typical work station.
4. Aisle space. Each operation must usually be serviced by an aisle. The portion of the aisle that is necessary to support the work station's activity should be included in the "footprint." Like the conveyor, if the aisle is

Equipment Square Footage

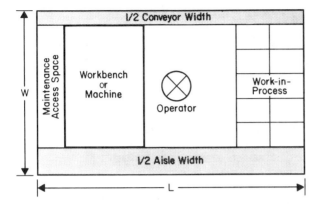

Fig. 7.1 Work station footprint.

usually shared, then only one-half the width of the aisle should be included.

5. Supplies. The footprint should also reflect space that must be allocated for supplies and supporting materials. The footprint should take into account such items as a tool chest, a supply cabinet, or processing materials such as oil, glue, or paint.
6. Maintenance access. In the event the equipment breaks down and must be dismantled, some access may be

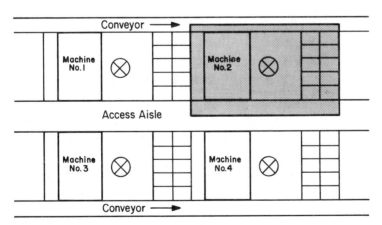

Fig. 7.2 Footprint within a layout.

required to adequately service the equipment. Room may be needed all around the equipment, not just that space utilized by the operator. Also, consideration must be given to the space needed for the removal for any of the equipment's components. For example, room must be available to remove a long shaft, or, perhaps, to bring in a crane. Space for maintenance access is critical, yet often overlooked.
7. Support equipment. There are many operations that require secondary support equipment. For instance, when wood is sawed or sanded, a dust collector may be necessary. Typically, a dust collector may be capable of handling multiple work stations. The footprint for these work stations should include a proportional share of the space required by the dust collector. Other examples of support equipment might include pollution control, environmental conditioning, and services/utilities (i.e., compressors).

When evaluating the square footage requirements of a work station, the planner should avoid overkill. A rough sketch of each operation's footprint is usually sufficient. The task of documenting a footprint is a case where time can be wasted on too much detail and the result would not be significantly improved. A rough sketch can be generated expediently and can be just as accurate.

Collecting data on space requirements can be handled much the same as collecting labor and equipment data. Information, in this case a space profile, must be developed for each operation within the manufacturing process. A sketch of the footprint should be made on the operation's data sheets which accompany the flow chart.

The availability of a footprint sketch for each operation will facilitate the design of future plant layouts. In addition to providing an equipment space template, these footprints also provide graphic documentation for numerical space projections.

When establishing a footprint, the footprint should be based on the square footage required, not necessarily the square footage that is used. There is no sense in planning future space requirements based on current space inefficiencies. An operation will tend to utilize the space available, rather than the space that is needed. Conversely, in a crowded factory, an operation may be confined to less space than it actually needs. Although, at first glance, a factory's space may appear to be efficiently utilized, it may cause an inefficient operation. For example, in a crowded factory, operators may have to leave their work stations in order to retrieve materials from a distant site. A footprint based on what is really required will ensure that future layouts have room at the work stations for materials.

Without intimate knowledge of each operation, the planner may be hard pressed to produce these footprints. Most likely, the planner will require assistance from manufacturing engineering, the production supervisors, or, perhaps, the operators themsleves.

Perhaps the most useful approach is to arrange for a walking tour through the factory with each production supervisor. Utilizing the process flow chart, the planner can stop the tour at each step and discuss the space needed at that step. A footprint can then be sketched based on the supervisor's knowledge and the planner's objectivity. This cooperative approach will also help to ensure that future plant layouts or square footage projections are not controversial. At least the base data—square feet per operation—will be based on extensive consultation with the production department.

A quarter-inch plant layout may also be helpful during this data collection phase. Rather than sketch out a footprint from scratch, the planner may simply mark up a layout. A layout may also be helpful when there are a number of pieces of the same equipment in one area. Assume that they all share a common storage area or common support equipment: then the square footage per piece could be "roughed out." The whole area could be divided by the number of pieces within that area. Depending on the nature of the manufacturing plan, a crude approach may be sufficient. Care must be taken to ensure that the designated total area includes only the space that is needed.

When quarter-inch layouts are utilized to block out footprints, there is a tendency to account for all the space. The planner should avoid trying to account for all the space since in any factory it is inevitable that 100% of the space will not utilized. Doors, I-beams, utility drops, etc., all contribute to space that cannot be effectively utilized. It is not necessary that one work station's square footage footprint border the next work station's footprint. It is also possible in a crowded facility to have footprints overlap when mapped out on a layout.

7.3 WORK-IN-PROCESS

As the product proceeds through the manufacturing process, there may be stops along the way. The major stopping points will show up as delays or storage steps on the process flow chart. Other stopping points may be hidden in the transportation steps. One common effect of these stops is to take up space on the manufacturing floor. The manufacturing plan must take into account the space required for these stops.

The data related to major delays and storages should be recorded on the operation's data sheet. Instead of space per piece of equipment, the planner should record the space per unit of product or material to be stored. It is also absolutely essential to record the length of time that it is to be stored. For example, a major delay occurs when a product must dry after it has been painted. The product utilizes factory floor space for the time it takes to dry. The drying time could have a significant impact on space requirements. The precise impact will depend on the length of the dry time, the size of the product, and the production rate of the product.

Major areas which are designated for work-in-process should be considered direct manufacturing space. As direct space, it should be included in any tabulation of utilized factory space.

7.4 INDIRECT AREAS

Much of a factory's floor space is not directly involved in manufacturing. When factory space is planned, these indirect areas must be taken into account. Indirect areas include the space required for major aisles, offices, break areas, restrooms, support equipment, supplies, tool crib, etc.

The indirect areas can be classified into three categories: employee-related areas, process-related areas, and major aisle space, which is related to the process, the people, and the facility itself. The techniques used for planning and, therefore, the data that must be collected depend on the category to which an area belongs.

Indirect Areas 115

7.4.1 Major Aisles

Major aisles account for the single biggest allocation of indirect space. The space required for the major aisles is a function of the quantity and type of traffic and also the design of the facility itself, The quantity and type of traffic could be either people or materials movement. In some cases, major aisles are needed for large numbers of people to move to and from the work areas. In other cases, such as in a highly automated facility, a major aisle may be needed to transport materials. Also, a major aisle must be available for the movement of the process equipment itself.
   The design of the facility will influence the percentage of space allocated to the aisles. For instance, a long narrow facility, such as an old mill building, will have a major aisle running down the middle of each floor. In such a facility, that aisle may only service a small amount of direct manufacturing space to either side. In a square facility, an aisle of the same width may service much more direct manufacturing space. When a projection of manufacturing space is being developed, the major aisles are of significant concern. In an unknown facility or layout, the major aisles can be factored as a percentage of the direct manufacturing space. In a known facility, such as the old mill, the aisle space is predetermined. In either case, the space allocated to major aisles is a significant and necessary component of any production operation.

7.4.2 Employee Support Areas

Every factory has space dedicated to meet the needs of the employees, such as break areas, locker rooms, and bathrooms. In these employee support areas, the space required is entirely dependent on the number of employees to be supported. It is relatively easy to determine if an area falls into this category. Ask the question, "What would happen to this indirect area if the entire work force were replaced by robots?" If the area would become obsolete, it is obviously employee related. Therefore, when future space requirements are planned for these areas, there is a direct relationship to the number of workers.

During this data collection phase of the planning process, that relationship must be defined.

The relationship between the number of employees and the support space required can be defined. It is usually necessary to combine knowledge of one's manufacturing operation with accepted industry practice. While there are some guidelines that can be followed, one must consider local building codes and the specific needs of the operation being planned. For instance, a noisy factory environment may necessitate on enclosed break area. A dirty process may require additional wash basins or perhaps showers. As in planning production areas, planning indirect areas should be based on the optimal space required, not necessarily the current space allocated. However, during this data collection phase, the planner should note current conditions. It may be important to have a record of the current space allocated for indirect functions. In addition, a note should be made of the capacity of those functions. Support space allocations should be expressed in terms of square feet per employee. Although each company is different, the following guidelines may be useful when employee-related space requirements are being developed:

1. Lavatory space—approximately 5 sq ft per employee.

    Toilets: 1 for every 20 men; 1 for every 12 women
    Urinals: 1 for every 35 men
    Wash basins: 1 for every 35 men; 1 for every 16 women
    Wash fountains: 1 for every 80 men; 1 for every 50 women

2. Supervisor's office—100 to 150 sq ft per supervisor. A determination must be made of the number of supervisors per employee. The number varies, depending on the type of industry, and may be documented company policy. The planner should consult with production management in order to establish a figure for planning purposes.
3. Break areas—8 sq ft per person.
4. Cafeteria—17 sq ft per person plus 22 to 35% for kitchen.
5. First aid area—1/2 sq ft per worker.

Indirect Areas

7.4.3 Process-Related Areas

Within the factory, there are indirect areas that are necessary to support the production process. Instead of being related to employee head count, these areas correlate to the production equipment or the factory space. Indirect process support areas are in addition to the direct storage and supply areas that are identified on the process flow chart. Some typical indirect process-related areas follow:

1. Tool crib, machine shop, and maintenance departments. These areas relate to the process itself. Some operations require a great deal of tooling and maintenance support, while others require very little. For planning purposes, it should be sufficient to project their space requirements as a ratio of the total production space. However, an effort should be made to ensure that the ratio is based on optimal space utilization, and not on current inefficiencies. The space currently needed for these support areas should be divided by the production space that is currently needed. The space that is needed may not be the same as the space that is utilized.
2. Utilities. Adequate space for furnaces, compressors, major transformers, or other support services must be incorporated. Utility space can be handled much the same as any process support area. A ratio must be established between the area required by the utility operations and the total production area.

7.4.4 Space Utilization Factor

A space utilization factor can be calculated by dividing the required factory space by the actual space utilized. Required space is the sum of the space required of the individual operations (the footprints) plus the space required of the indirect functions. The actual space utilized is the size of the factory minus empty space such as that set aside for expansion.

Most plants, unless severely overcrowded, will not show 100% utilization. There will always be space that cannot be used due to physical limitations. For example, for every column, or I-beam, 9 to 16 sq ft are lost from productive floor space. Other limitations can be attributed to stairways, doors, utility lines, trenches, etc. These factors can create "dead space" by limiting plant layout flexibility. Some dead space is inevitable and must be planned for.

Typically, the space that remains unaccounted for amounts to roughly 5% of the factory space available. Therefore 95% space utilization is usually a reasonable figure to use for planning. Any factory with a significantly higher or lower space utilization percentage usually has other problems that must be addressed. These problems should not be included in future plans, unless primarily attributed to physical limitations that will not change. Tight column spacing, such as found in old mill buildings, is a good example of fixed physical limitation.

7.5 OFFICE SPACE

Office space may be required within the manufacturing facility for administrative, technical, and clerical personnel. In small companies, there may be only one facility that houses the entire company including manufacturing. In any case, the manufacturing plan should account for the office space required within the facility.

Office areas can either be ratioed to the manufacturing area required or be based on a projected head count. The approach used will depend on the availability of the information and the level of detail required of the plan.

7.5.1 Office Space Ratios

To establish a crude estimate of future office space, it may be sufficient to utilize a ratio of office space to factory space. What is required for this ratio approach is simply a record of the current space allocation. From this information the square footage utilized for office areas can be divided by the square footage required for direct manufacturing. From these data, it is easy to develop a ratio of office space to manufacturing space. Once the future manufacturing space is calculated, this ratio can be used to calculate the office space required. The ratio approach

Office Space

assumes that office space is simply a percentage of the manufacturing space. Also assumed is the long-range validity of this somewhat questionable relationship.

In reality, the office space required for various business functions may have little or no relationship to the space required for production. For example, future factory space may be reduced or at least not expanded through improved productivity. At the same time, office space may increase because of increased emphasis on customer service, marketing, research and development, and information management. While the ratio approach may be good for a quick and dirty estimate, the planner must recognize its inherent faults and limitations.

7.5.2 Office Space from Manpower Projections

Administrative, technical, and clerical manpower planning are beyond the scope of this text. They are also outside the discipline of manufacturing planning. However, it is not beyond the planner's role to convert manpower projections into office space requirements.

Departmental staffing levels should be forecasted, and by the management of each department. Once approved by upper management, the manpower forecast can be used by the planner to calculate the space required for offices. The manpower forecast may simply state the department name and the number of people. It is then up to the planner to accurately project the office space required.

In a typical office environment, approximately 200 sq ft per person is needed. In addition to the office itself, this figure takes into account support areas as illustrated in Table 7.1. Of course, the planner may wish to modify these figures to take into account the nature of one's business and the specific needs of each department. For instance, engineers in product development may require space for the storage of samples and prototypes. Another example can be seen in a dedicated factory staffed only by production management. The space required for conferences, receptions, and laboratory facilities may be minimized.

The square footage required per person may also be constructed or checked by dividing a department's space allocation by the number of people employed within that department. This simple ratio approach may be especially helpful when a department

Table 7.1 Office Square Feet per Employee

Space per employee	Square feet
Average office size (9 × 10)	90
Minor aisle	20
Conference rooms	3
Supply rooms	5
Common area (reception, photocopier, mail slots)	6
Computer rooms	5
Cafeteria	6
Lavatories	5
Major aisles/corridors/stairs	20
Facilities services (heat, phone, electric)	20
Facilities constraints (walls, columns)	15
Miscellaneous (medical, library, security, etc.)	5
Total	200

with unusual space requirements is evaluated. The space required per person for research and development, testing laboratories, product design, or even marketing offices may be unique to the company. In advance of the manpower forecast, the planner should calculate the space required per person for each area.

Sometimes a manpower projection may be more detailed. The planner might be provided with a breakdown of each department's manpower by staff level. The manpower forecast may state the number of senior executives, department heads, supervisors, and staff personnel. The size of an office usually varies with management level. A senior executive will require 300 to 400 sq ft for his or her office, while a department head will require 150 to 300 sq ft. A supervisor will require only 100 to 200 sq ft and staff personnel 75 to 100 sq ft.

Office Space

The planner should evaluate each office area that is to be housed in the manufacturing facility. Prior to developing the base data, the planner must also understand the level of detail which will be provided in the manpower projections. The planner can then establish the square feet required per person to coincide with the level of detail delineated in the manpower forecast.

8
Productivity Factors

8.1 INTRODUCTION

Manufacturing is a highly complex interaction between men, materials, and equipment. To facilitate understanding and control of manufacturing, most of these interactions are analyzed and quantified. In one form, these interactions are quantified by direct labor standards and machine cycle times. While these figures may represent a realistic hourly capacity for a production operation, sustaining these rates for the long term is merely theoretical. There are a number of factors that will significantly alter the theoretical long-term capacity of each operation. Proper understanding and utilization of these productivity factors will enable the development of a realistic and practical manufacturing plan. Productivity factors include performance, utilization, uptime, attendance, yields, rework, and improvement factors. Productivity factors serve two purposes: (1) they define the inevitable inefficiencies of the operation, and (2) they rectify inaccuracies in the original definition of the standard or cycle time. In either case, the productivity factors used will have a dramatic effect on future manufacturing requirements. For instance, these factors may determine up to 50% of the direct labor required. Due to this large impact, productivity factors deserve serious attention.

Effort must be spent on both defining each factor and then quantifying each factor. Since there are many variations of manufacturing controls and reporting systems, the definition of the productivity factors may also vary among companies. Both the definition and collection of productivity data may lead to some controversy. However, it is better to settle any dispute at an early stage rather than later when the plan is completed. These disputes usually involve a conflict between the negative appearance of a low productivity factor and the need to be realistic. To minimize this controversy, the planner must be prepared to present and defend the reasoning that went into each factor.

One method of minimizing this conflict is to establish long-range goals for each factor. Each year of the manufacturing plan can be based on an improvement in the productivity factors. Productivity factors then become variable depending on the year being studied. These variable factors should not significantly impact the difficulty of planning. In fact, incorporating these factors as a variable could significantly minimize controversy and maximize accuracy.

8.2 DEFINITION AND IMPACT ON MANUFACTURING PLANNING

Productivity factors are used to measure the effectiveness of the manufacturing operation compared to equipment and labor standard data. An accurate manufacturing plan must take into account inaccuracies in the standard data and inefficiencies in the manufacturing process. The productivity factors defined below provide the means for taking into account the discrepancies between theoretical and realistic capacity.

8.2.1 Performance

Performance is a measure of labor effectiveness compared to an expected standard. For the time spent on a measured task,

Definition and Impact on Manufacturing Planning 125

performance is the actual output divided by the expected output. For example, assume an engineered labor standard for a job was 0.01 hours per unit. This standard time converts to a standard production rate of 100 units per hour. Also assume that an employee works solely on that job for an hour and produces only 80 units. Eighty units divided by the expected 100 units yields a performance factor of 80%. Similarly, if the employee had produced 120 units, the performance would have been 120% of standard.

In measuring performance, the key point to keep in mind is that it is a measure of on-the-job performance. It should not be used to account for time spent away from the job, which is not covered in the standard itself. The only exception is when an elemental task that is essential to the job has been inadvertently omitted from the standard. This omission would then be reflected in the performance factor.

Performance can vary greatly between industries and companies. In a low-key, day-work environment, employee performance will typically run under 100%, whereas in a high-pressure, piece-work incentive environment, performance could run well over 100%. Incentives play a major part in performance. This is not to suggest that piecework incentives should be implemented in every environment. While piecework incentives may enhance performance, they do not necessarily enhance productivity. There are trade-offs. A high performance at any cost could lead to inferior quality, excessive rejects and rework, absenteeism, hoarding, and cumbersome labor reporting.

In addition to labor effectiveness, the performance factor has the effect of correcting inaccuracies in the labor standard. In the earlier example, 100 units per hour was the standard expected production rate. If the standard had been done accurately, perhaps only 90 units per hour should have been expected. The actual performance of 80% could mean that 10 of the 100 units per hour should not have been expected at all and that another 10 units were lost due to a lack of incentive.

Whatever the cause, this performance factor must be taken into account in the manufacturing plan. Employee performance will affect both the manpower requirements and the equipment requirements. In many cases, the effective capacity of a piece of equipment is controlled by the operator. Likewise, an operator's production may be limited by the cycle time of the equipment. In a situation where the operator and the equipment are intertwined, performance is a reflection of the operation as a whole. In a manufacturing plan, performance is a necessary and realistic consideration when developing both manpower and equipment requirements.

Although performance may vary between operations, an average departmental performance may be sufficient for manufacturing planning. Assume that labor standards are set the same way throughout the factory. In an environment of uniform incentives, the performance levels should fall within a narrow range. However, there may be situations where the standard is way off base for a particular operation. An incorrect standard will show up in the performance factor for that operation. While this situation may have a negligible impact on the overall department's head count, it could significantly impact the operation's equipment performance.

In an environment where there are no engineering standards available, a reasonable expectancy time or estimated standards might be used. If the estimated standard is based on actual production, performance may already be included in that standard. Therefore, there may be no performance information available or even necessary in this situation.

8.2.2 Utilization

Direct labor utilization is the percentage of factory labor which is directly applied to production. It refers to only that time spent adding value to the product, otherwise known as applied labor. Applied labor refers to that time spent on productive work regardless of performance. In the earlier example, one worker produced 80 units per hour, while the other worker produced 120 units per hour. While their performance varied, an hour was spent adding value to the product. In either case, there was 1 hour spent of applied labor.

Utilization is calculated by dividing the applied labor by the sum of the applied plus the unapplied labor. Unapplied labor is the time spent by factory labor in nonproductive work. Unapplied labor does not add any value to the product; while it is unproductive, it is also unavoidable.

Unapplied labor takes into account a variety of activities. These indirect activities are listed below:

    Inspection
    Training

Definition and Impact on Manufacturing Planning 127

Material handling
Clerical and administrative—in support of direct labor operation
Lead man—time spent in setting up and training others
Physical inventory
Housekeeping—general upkeep of the production area
Rework—caused by the production department
Rework—created by engineering changes
Rework—caused by defective parts
Rework—created by another department
Lot screening—screening rejected parts for defects
Maintenance of equipment
Business meetings
United Fund drive
Bloodmobile
Personnel consultations
Medical—first aid
Medical—health insurance, such as changing bandages or taking medication
Downtime—time lost due to power failure or machine repair
Waiting time—due to parts shortage

Primarily, unapplied labor should take into account that time that is spent at work but not on a productive job; it should include the time spent by factory workers only, not supervisors or engineers. In many companies, factory workers are delineated by those within the production departments who are paid hourly.

8.2.3 Attendance

Attendance is calculated by dividing the days per year worked by the days per year available. The attendance factor is used to factor in employee absence when calculating future manpower requirements.

Total working days available are the number of days per year that the factory is open and operational. They should not include holidays or scheduled shutdowns. For instance, there will be 240 working days available in a factory running 5 days/week, less 10 holidays and a 2-week shutdown.

Employees do not work every available working day. The average number of days worked per year per employee can best be figured by subtracting the number of days absent from the number of days available. Absence refers to only those days when employees are missing but the plant is open. Therefore, holidays and vacation time taken during a shutdown should not be counted as absence. Holidays and shutdowns are not available as working days. Absent days are the average number of working days missed per employee per year. Absent days include both paid and unpaid sick days, personal dyas, vacation days, and any other authorized or unauthorized leave of absence.

In the example above there are 240 working days per year. Assume that the average employee has 2 weeks vacation, but 1 week of that vacation is taken during shutdown and 1 week from working days. Also assume that the average employee calls in sick, or with car trouble, or with personal problems 1 day/month. In addition the average employee spends 1 day/year on jury duty. In this instance, the attendance factor would be 92.5%, as illustrated in Table 8.1.

A couple of points need to be reiterated. One is that time off must be counted whether it is paid or unpaid. Whatever the cause, the factory must be staffed to make up for the missing people. The second point is that the definition of attendance be very specific. Absent days must come only from days available.

8.2.4 Uptime

An uptime factor is the percentage of available time that an operation is capable of running. It is used only to determine an operation's equipment availability and therefore future requirements. It is similar to the utilization factor, which is solely a measure of labor. Like utilization, uptime can be calculated by subtracting the nonproductive time from the total time available. Typically, nonproductive equipment time is called downtime.

Downtime is usually used to describe time that is lost due to equipment failure. However, preventive maintenance, inspections, power failures, and minor malfunctions also contribute to downtime. For the purpose of manufacturing planning, downtime should include any other factors that contribute to an equipment's nonproductive time.

Equipment is nonproductive whenever it is not operating. Whether due to material shortages or a missing operator, if the lost time is inevitable and predictable for the foreseeable future, it should be included.

Definition and Impact on Manufacturing Planning

Table 8.1 Calculating Attendance

Days absent	Reason
5	Vacation (during working days)
6	Sick
5	Personal time
1	Car trouble
1	Jury duty
18	Total days absent

$$\frac{222 \text{ days attended per year}}{240 \text{ days available}} = 92.5\% \text{ Attendance}$$

In an extreme case, the operator's attendance and utilization may have to be included in an uptime factor. Although rare, assume a factory with work benches dedicated to each employee, much like desks in an office. If each work bench contained specialized production or test equipment, that equipment would not run without the operator present. If this situation is expected to continue, future work bench and equipment requirements must take this into account.

Production equipment may be also rendered nonproductive when the operator shuts it down to get more materials. This situation creates unnecessary but predictable downtime. Obviously, this situation should change. If it is realistic to think that it will be changed, it should not be used in the uptime figure used for manufacturing planning. If, however, it is realistic to believe that no change will occur, then this lost time should be planned on. A manufacturing plan should present a realistic projection of future requirements.

Many of the same factors that contribute to nonproductive labor utilization can also contribute to nonproductive equipment time. On some operations, the equipment may be shut down to allow the operator to obtain training, fill out forms, or, as previously mentioned, retrieve materials. In other operations, the equipment is kept running at all costs. Dedicated material handlers and clerical personnel enable the operator to maximize production by running the equipment. Therefore, equipment uptime will be maximized.

An uptime factor must be generated for each operation. There is usually no way to generalize uptime for a work center or an entire production department. For example, within the same factory, a work bench with a soldering iron may have a low uptime percentage, while an automatic component insertion machine may have a high uptime factor.

A key point to keep in mind about uptime is that it is distinct and separate from utilization. Uptime is a measure of equipment and utilization is a measure of labor. Both may share some of the same elements and yet not share others. A good example is when equipment is shut down for preventive maintenance. Properly planned, the operator can run a second piece of equipment while the first one is down. In this case the equipment is affected by lost time, while the operator is not.

8.2.5 Yield

Scrap losses and rework have a major impact on manufacturing requirements. It is imperative that the manufacturing plan be based on a realistic delineation of scrap losses and rework loops. Scrap losses and rework loops can occur throughout the manufacturing process. Where they occur should be documented on the process flow chart. The extent of the rework and the product loss must be quantified and some figures established for planning.

Yield is the percentage of product that is processed successfully. Yield is the output divided by the input whether for the entire manufacturing process or for a single operation. In general, the term yield refers to the percentage of product that is not scrapped or down graded. A 90% yield means that out of 100 units started, only 90 units were finished successfully.

Many manufacturers do not simply scrap inferior products. In many cases, products that do not meet quality standards are downgraded and sold as seconds. Some manufacturers may even establish a hierarchy of downgrade categories.

Yield losses must be factored into a manufacturing plan. For example, in a factory with a yield of 90%, it is necessary to start 111 units to successfully complete 100 units. In order to account for this loss, the production line must be capable of processing 11% more product at the beginning of the line than at the end of the line. Obviously, it then becomes important to

Definition and Impact on Manufacturing Planning 131

determine at what point in the process the yield loss occurs and the extent of that loss.

Rework can also have a significant impact on manufacturing requirements. Products that must be reworked create the need for repair areas, additional inspection capacity, and, perhaps, additional manufacturing capacity if the rework is done on the production line.

Products requiring rework may be classified into categories. For instance, a rejected product may be designated for cosmetic, electrical, or mechanical repair. A rejected product could follow three different paths depending on its defect. Assume that the repaired product is returned to the same inspection operation. The planner must then determine if this repair cycle is infinite or finite. That is, how many times can a product get re-inspected and re-rejected before it is scrapped? Also, does the product's second pass through inspection yield the same results as the first pass? It is usually imperative that the answers to these questions be reflected in the manufacturing plan.

Fig. 8.1 demonstrates the effect that both yield and rework rates can have on a production line. In this example, 100 electromechanical assemblies are started each day. Upon inspection, 10% are scrapped, 10% are sent to a cosmetic repair area, and another 10% are sent back to the mechanical assembly operation. Only 70% of the assemblies are successfully processed the first time.

The reject rate of 30% has a major impact on the manufacturing operation. This reject rate creates a load on the mechanical assembly operation of 111 units per day (100 initially, 10 for rework, and 1 for re-rework). The inspection operation must handle a quantity of 122 units per day as detailed in the example. In the end, the process yields only 88 units per day for the final packaging operation. In effect, this production line produces an 88% yield. To actually produce 100 units per day, 114 units would have to be started. The inspection operation would have to inspect 139 units. If these conditions were projected to continue, the inspection operation must be planned to handle 139% of the desired production volume.

8.2.6 Productivity Improvements

Most manufacturers strive to improve the productivity of their factories. The expected effects of this effort should be documented

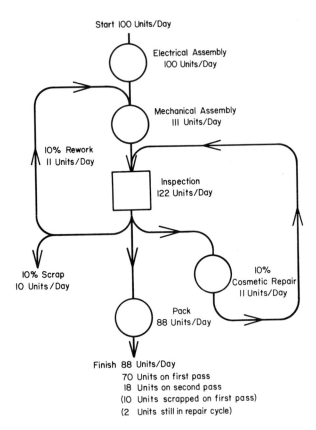

Fig. 8.1 Yield/rework impact.

within the manufacturing plan. The term productivity is used here in a narrow sense. A productivity improvement factor is used to approximate an expected reduction in standard hours, equipment cycle times, and process improvements. Expressed as a percentage, the productivity factor notes the results of the expected improvements. For example, using an annual 10% reduction in the standard hours required per product yields a productivity improvement factor of 72%. For every standard hour required today, only 0.72 hours would be needed 3 years from now.

The planner may wish to use a productivity improvement factor only on certain operations or work centers. Short- to medium-range manufacturing plans may require this more detailed approach. However, long-range manufacturing plans may require only a general improvement factor. It may be sufficient to utilize the same productivity improvement percentage for the entire process and product line.

The productivity improvement factor can also be used to reflect the impact of tightening standards that are currently loose. Assume that the existing direct labor standards are loose. If there is no incentive to produce (such as piecework), loose standards will enable the work force to produce what is expected with little or no effort. If these loose standards are recognized and documented, a productivity improvement factor can be utilized to take into account a future tightening of the standards. The planner must quantify the looseness of the standards and the pace with which they will be corrected. A productivity improvement factor can then reflect the extent and the pace of the projected improvements.

Productivity factors can sometimes be tricky. Each company (and sometimes each individual) may have its own definitions and methodology for including each of these factors. It is not important that the definitions and methods used for planning be the same as described here. However, it is important to accurately incorporate every legitimate and realistic impediment to productivity. This means not overlooking or overcompensating (such as double counting) for any of the elements that impact on operation's productivity.

8.3 COLLECTING THE DATA

Productivity factors will have a significant impact on the manufacturing plan. These factors must be based on the most reliable information available. Due to the significance of their impact, it is likely that the factors used will be scrutinized closely. It is therefore imperative that these factors be both realistic and auditible.

The data used to develop these productivity factors come from a variety of sources. These data sources can be broken down into three groups: reporting systems, production reports, and engineering studies.

8.3.1 Reporting Systems

Past and current data on many of the productivity factors can be extracted from a variety of computer-based reporting systems. Of particular interest are the systems that monitor and control factory labor and payroll. Less common but also in use are systems that monitor and report on equipment utilization, maintenance, and process activity.

Labor Reporting Systems

Labor reporting systems are used primarily to monitor direct labor activity on the production floor. These systems usually record the daily productivity of each worker. At the end of each day, a worker will note the job worked on, the units produced, and the length of time that each job took. Nonproductive jobs such as housekeeping and material handling are usually noted also. At the end of each day, these data are collected and entered into the labor reporting system.

Many of these systems will compile all of the data and generate a summary report. Many summary reports will present the performance factor for each job operation or work center. For the quantity of product produced, the report may present the expected standard hours versus the actual hours. The report may also present a breakdown of the nonproductive hours. Such a breakdown will allow calculation of the direct labor utilization factor if it is not already presented.

Although not recommended, a labor reporting system may be helpful when establishing uptime. Assuming that the nonproductive direct labor time is categorized, the summary report may present the time spent waiting for materials and waiting for equipment repair. However, the planner should treat this information cautiously and in most cases avoid it altogether. In a properly managed factory, no worker should spend too much time waiting for equipment to be repaired or materials to become available. In either of these situations, the worker should be placed on a new job. Therefore, the equipment will still be idle while the worker is once again productive.

Collecting the Data

Payroll Systems

Although payroll systems may prove to be inaccessible directly, the planner may have access to its summary reports. Payroll systems are of interest to the planner since they typically monitor vacations and attendance.

Vacations are pertinent since vacation time is a necessary component of the attendance factor. An older established manufacturing department may average more vacation time than a department with new hires. The planner should also be interested in when these vacations are taken. Vacation time taken during a shutdown is of little consequence.

A payroll system should also highlight the number of days lost due to sickness or other personal reasons. There is one caution when attendance is calculated from a payroll report. Unpaid time off may not be adequately documented since there is no impact on payroll dollars. Unpaid time off may or may not be displayed, depending on the system.

Maintenance Systems

Some plants have computer-based maintenance reporting systems. These systems are used to monitor the activity of the plant's maintenance workers. Like the labor reporting system, this system is used to record the jobs done and how long they took.

An evaluation of the time spent on maintaining production equipment, both planned and unplanned, will allow pertinent downtime data to be established. The planner must be careful to attribute to downtime only the time that production equipment is down during productive time. Equipment that is maintained or repaired during an offshift or on a weekend does not hamper production. Therefore, equipment uptime is not affected.

Process Monitoring Systems

Some factories, particularly processing plants, have computer based systems that monitor processing activity. It is difficult to generalize about these systems. They are usually tailored to the industry of even the company where they are installed. These systems may monitor the output quality of each processing step as well as the rate of that processing. A summary of this information could be helpful in determining an operation's performance, uptime, and yield.

8.3.2 Production Reports

In most factories, production management must periodically generate manufacturing status reports. These reports are usually tailored to the requests of senior management. It is difficult to generalize on their content or format. However, the planner should become familiar with the reports that are produced and the information they contain. These reports may be especially helpful if they are generated on a routine basis. The availability and content of these reports may enable the planner to accurately quantify some of the productivity factors. There may be weekly reports that document the results of the inspection operations. All of the data necessary to accurately develop the yield factors may be clearly spelled out. Not only would the planner have a picture of the current scrap losses, down-grades, and rework percentages, but also a look at the trends.

In some environments, there may be a weekly equipment utilization report. This report clearly details the utilization of each major piece of capital equipment. The report may present the equipment's schedules hours versus actual hours of run time. Next to any discrepancy in those hours, there is usually a notation as to the reason. On such a report equipment uptime can be calculated for all the major pieces of production equipment.

Production reports may also highlight the effectiveness of the work force. These reports may contain information that is helpful when developing the utilization factor. These reports may be include an overview of how and where direct labor has been used. In addition, such reports may contain data on overtime, staffing levels, and, perhaps, attendance.

These reports by production management are done to support the desires of senior management. Perhaps the best way to uncover these reports is through a discussion with each production manager. The availability and content of these reports vary greatly between departments. It is up to the planner to determine what is available and what is useful.

8.3.3 Engineering Studies

Both manufacturing engineering studies and industrial engineering method studies contain data useful for developing productivity factors. Similar to the reports requested by production management, engineering studies are company specific although, unlike production management reports, engineering studies are usually done on a random basis.

Of particular interest to the manufacturing planner are any studies that involve the effectiveness of present and proposed manufacturing alternatives. The planner should be concerned with those studies that provide information in support of a productivity factor. The planner should also be concerned with engineering projects that are likely to generate significant improvements in productivity. General information on performance, utilization, attendance, uptime, and yield may all be available from other sources. However, at the operation level, manufacturing/production engineers may be the only source for productivity improvement data. It is their studies which will quantify the future productivity potential for a given operation.

Manufacturing engineers typically generate studies regarding an equipment's productivity potential. These studies may detail the effects of new feeder mechanisms, alternative materials, automated equipment, or any technique or technology that will reduce processing time and increase quality. The data contained in some of these studies may be of particular interest to the planner.

Detailed manufacturing engineering studies may also be conducted to uncover yield problems. These studies may contain the yield data required for manufacturing planning. Uncovering such a study could eliminate the need for the planner to duplicate this effort. Not only might the current yield situation be accurately documented, but also the potential for improving that yield.

Manufacturing engineers may also be asked to investigate equipment downtime. Looking for a cause and solution for a downtime problem, the engineer may very accurately document the downtime itself. Even if the engineer corrects part of the problem, this documentation will provide a downtime figure for planning purposes.

Manufacturing engineering studies tend to focus on individual operations within the manufacturing process. Many industrial engineering studies, however, focus on the interrelationship between these operations. The studies that focus on this interrelationship may be categorized under material handling analysis, method studies, or line balancing. Any of these studies may contain documented information on uptime, yield, rework, as well as labor utilization and performance. Any one of these factors could come under scrutiny in an industrial engineer's study of the manufacturing operation. For instance, the feasibility of an automated material handling system may depend on improved direct labor utilization. Such a study may document both the time spent handling materials and the total nonproductive time. The study may also project handling time that could be reduced with the introduction of an automated system. From this study, both present and future direct labor utilization figures could be extracted. Also, handling times that are included in the labor standard may be reduced. The extent of that reduction could be reflected in the productivity improvement factor.

Most of these manufacturing/industrial engineering studies are unlikely to surface on their own. The planner must make a concerted effort to uncover them. Before requesting and evaluating engineering studies, the planner should have a clear idea of what information is needed. In most environments, productivity factors can be collected from routine reporting systems and production reports. If the information is unavailable from these routine sources, the planner should consider uncovering the data from an engineering study.

8.4 PROJECTING IMPROVEMENTS

Productivity factors must be considered a variable. Usually they are not expected to continue their current status for years to come. Most manufacturers work very hard at improving these productivity factors. The expected results of this continued effort should be reflected in the manufacturing plan.

8.4.1 Production Management Goals

There are many industries where some or all productivity factors are monitored very closely. In such a setting it is not unusual for an improvement goal to be established for each of these factors. Production management and the supporting engineering groups are responsible for attaining these goals. Some of these goals may be expressed in general terms. For instance, management may establish a goal for improving employee attendance. In this instance, the goal for the attendance factor is applicable for planning the entire direct labor force. In other situations,

Projecting Improvements

the goal for these factors may be very specific. Goals may be established to rectify specific yield problems on specific products. There may be uptime (or equipment utilization) goals independently established for each major piece of equipment.

Whether a goal is established factory wide or by work center or for specific operation, the manufacturing plan should utilize the detail provided. Properly constructed, the planning process should be able to handle any of these factors at any level of detail.

Typically, the management goals for these productivity factors will not be expressed in the terms desired for manufacturing planning. The planner may have to convert management goals into planning terminology. For example, management may express the desire to reduce absenteeism by 10% per year for the next 3 years. This desire must then be calculated into manufacturing plan's definition of attendance.

Production management goals for the productivity factors must be expressed in planning terminology. They also should be noted separately for each of the time periods being studied. A record of the productivity factors used for each year of fiscal quarter will allow for easy reference and changes later.

Every effort should be made to use goals that have already been established, assuming, of course, that the goals are realistic and achievable. If so, there is no reason why the manufacturing plan should not be based on the achievement of these goals rather than on current conditions.

8.4.2 Engineering Cost Reduction Goals

Engineering projects may clearly define expected improvements in the productivity factors. Like production management goals, they should be utilized in the manufacturing plan.

Engineering studies may very accurately quantify past and present operating conditions and therefore productivity factors. These same studies may also demonstrate the potential for improvement in these factors. If the study has been accepted and a project is underway to make these improvements, then a goal has been established. Assuming that the goal is realistic and significant, its impact on the productivity factors for each time period should be calculated.

8.4.3 Historical Trend Analysis

Some factories may operate with no stated goals for improvement of productivity factors, yet in one form or another, the information is probably available. The planner could develop productivity factors based on the current operation.

Unfortunately, the productivity factors that represent the current operation may not accurately reflect the future operation. Due to the significance of these factors, it may be necessary for the planner to utilize trends. For instance, a factory that is becoming more tightly controlled may show a consistent improvement in direct labor utilization year after year. Conversely, another industrial plant may exhibit a decline in some of the productivity factors.

Productivity factors, whether improving or declining, should reflect the results of these trends. Of course, some of these trends may not be linear. Factors that are improving may be the result of a learning curve and as time passes the improvements become less significant. The learning curve effect on productivity factors can be seen when a new product is introduced. During the first few years, both performance and yield may improve significantly. Eventually, the bugs are worked out and the pace of that improvement decreases.

Most likely, there will not be enough data or time to do sophisticated trend analysis. If one is lucky, some historical data on the productivity factors may be recorded for each of the last 2 to 5 years. Unfortunately, two to five points on a curve are insufficient for sophisticated trend analysis. Even if the data were available for each week or month during the last 2 to 5 years, sophisticated analysis could be a waste of time: a minor improvement in accuracy would not offset the major commitment in time.

Through crude, but helpful, historical data, the planner can simply eyeball any trend's impact on future productivity factors. The key to the success of this approach is to review the productivity factors with management before they are used in the plan. Management is then forced to approve a major input into the manufacturing plan. It is an opportunity to get them involved in the manufacturing planning process and instill their confidence in it. Besides, these trends that are eyeballed to require judgment as to their validity. Because of the somewhat subjective nature of this approach, the productivity factors that are used should be agreed upon by management.

9
The Planning Process

9.1 INTRODUCTION

The actual task of calculating the manufacturing requirements can be tedious. The mathematics are simple, but the number of the individual calculations may be large. Through clear direction and organization, the planning process will become a straightforward procedure.

By placing the planning process in perspective, one can readily see how it must be organized. The calculations must be done and documented in such a fashion as to enhance the plan's flexibility and auditability. These objectives are best met through understanding the distinction between requirements planning and tactical planning. Equally important to the plan is the physical layout of both the data and the calculations. The data and calculations must be presentable whether the calculations are performed manually or via computer. The planning process can also be made easier by following some rules of thumb. There are some general ground rules that will provide direction for many of the minor decisions that must be made during the planning process.

9.2 REQUIREMENTS PLANNING VERSUS TACTICAL PLANNING

Manufacturing planning may be described as the process of determining the manpower, equipment, and facilities required to meet forecasted production requirements in the most timely and cost-effective manner. A manufacturing assets plan is a two-part process involving the determination of future manufacturing resource requirements and the tactical approach for meeting those requirements.

A requirements plan is the straightforward result of calculations based on established parameters. The development of a requirements plan is objective in nature. It does not require decision making, since the operating parameters and factors are defined during the data collection phase. However, some of these "given" data are based on subjective decisions. These decisions may have to be reconsidered once the first pass at a requirements plan is available for review. For instance, future equipment requirements can be expressed in hours per week required. There are a number of ways to obtain those requirements. Decisions on the number of shifts, overtime, and the affordability of new equipment will all impact the actual equipment units required. Even in a projection of the hours required, some tactical decisions are included. Perhaps most significant are those decisions that involve items that will be produced in-house versus items that will be offloaded to vendors.

In practical application, requirements planning and tactical planning are intertwined. This does not mean that the two are indistinguishable. Requirements planning is a feedback mechanism to tactical planning. Data based on tactical decisions are input to the requirements plan. Based on the resulting projection of manufacturing requirements, the tactical data may be altered and a new requirements plan generated.

Typically, a preliminary manufacturing plan may be based on company policy; e.g., two shifts per day, 5 days/week, and no overtime. Subsequent modification of the plan may take place due to the economics of operating on a thrid shift rather than purchasing new equipment. Such a decision is tactical since it involves determining the best way to obtain the equipment hours required. Similarly, it involves tactical planning to determine at what plant site the manufacturing should be done or what should be offloaded to outside vendors. Any of these tactical decisions must be based on a number of parameters. A projection of manufacturing resource requirements is only one of the parameters that must be considered. Other considerations might include taxes, labor rates, access to raw materials and markets, quality, and management. However, many of these factors are outside the responsibility of the manufacturing planner.

Distinguishing between requirements planning and tactical planning will clarify one's direction. The planner will more readily discern when it is necessary to consult management during the planning process. Requirements planning involves number crunching, while tactical planning involves decision making. Therefore, the requirements planning process should be organized to facilitate its role as a feedback mechanism to the tactical planning process.

To reiterate, it is important to understand that in the overall context of manufacturing assets planning, there are two distinct phases: (1) the requirements planning phase and (2) the tactical planning phase. Although requirements and tactics are intertwined, the two phases can be delineated. Requirements planning is the process of determining a preliminary manufacturing plan based on predefined parameters. Requirements planning ends with the preliminary projection of manufacturing requirements. Manufacturing planning becomes tactical when those parameters are altered as a result of reviewing the preliminary requirements plan. The objective of the tactical planning phase is to optimize the viability of the requirements plan through manipulation of the operating parameters.

9.3 CALCULATING REQUIREMENTS

Once the data have been collected and organized, the process of calculating manufacturing requirements can begin. Like the data itself, these calculations must be well organized and documented.

Documentation of the calculations involved in the manufacturing plan serves the same purpose as documenting the data. Organization and thorough backup allow for inquiry into the procedures and assumptions used. A planning process that is auditable becomes more credible. Documentation will allow last minute changes to become more readily incorporated. Errors can become more easily identified and corrected. Documentation will enhance the planning system's flexibility by ensuring the capability to rapidly generate a new manufacturing plan based on new parameters.

Thorough documentation, especially within the final report, may be useful as a reference for other studies. In a feasibility study for an automated materials handling system, the manufacturing plan may be especially helpful. A well-documented plan may need to be consulted to determine how many production machines the automated system must support in the future. Also, the manufacturing plan may need to be audited in detail to determine if the handling system itself will reduce the number of machines required. Only through adequate documentation can the manufacturing plan live up to its potential as a corporate reference document.

The planning calculations must be systematized. Specifically, the numerous calculations will require the utilization of standard accounting work or spread sheets, although a more efficient alternative may be to utilize a spread sheet software package. (Spread sheet programs will be discussed in Chapter 14.) In either case, the approach to calculating requirements must be consistent throughout the planning process. The same format should apply to every production line or manufacturing operation that is being planned.

In some manufacturing environments, a practical way to handle these calculations is through the use of a computer. The primary determinant is the level of detail desired. A detailed manufacturing plan may be possible only through a computerized approach. A computer-aided approach may be especially necessary in a multiproduct, multicomponent, highly complex manufacturing industry such as electronics. A major consideration is the frequency with which the manufacturing plan must be generated or regenerated.

Whether designing a program or calculating by hand, the same general ground rules apply when a requirements plan is being generated. These rules of thumb provide direction for many of the minor but numerous decisions that must be made during this calculation phase. Some of these rules are the same as those that apply to the data collection phase, only their application is slightly different:

1. Avoid excessive detail. For example, when calculating requirements, utilize only the number of decimal places that is really needed. An extra decimal place may add significantly to the effort required, yet result in an insignificant improvement in accuracy.
2. Err on the side of excess capacity. For manufacturing planning, playing it safe means ensuring that future

Determining Strategy 145

capacity is available. Any round-off errors, or other borderline situations, should be slanted in favor of ensuring sufficient capacity.
3. Neatness counts. Organization and documentation of these calculations will help to ensure that the plan is accurate, credible, auditable, and revisable.

9.4 DETERMINING STRATEGY

While requirements are calculated objectively, strategy is conceived subjectively. Strategic decisions are based on qualitative as well as quantitative information. This background information as well as the strategic decision itself must be documented. Recording this information is done for the same reason ad documenting the data and calculations involved in manufacturing planning.

The strategic planning phase, due to its more subjective nature, has its own set of ground rules. These ground rules relate more to the planner's attitude than the planning technique, although an effective technique or approach may result from a proper attitude. Alternative operating scenarios can best be evaluated by adopting the following strategic planning guidelines:

1. Be flexible. An open mind must be maintained. Consideration should be given to alternative strategies whether conceived by the planner or others. Keeping flexible means giving consideration to any reasonable alternative strategy. Alternatives could include the number of shifts, manufacturing site, offloads, vertical/horizontal integration of manufacturing, etc.
2. Be innovative. Beyond keeping an open mind to alternative strategies, the planner should actively seek creative alternatives. Opportunities are especially prevalent when postulating new manufacturing facilities.
3. Think long range. If the plan is long range, then think in those terms. Avoid interjecting short-range, day-to-day problems into a long-range planning process, yet keep in mind that a long-range plan must be achieved in stages. The foundation of the plan still rests in the realities of the present.
4. Consult others. Strategic planning implies decision making. Any decision of strategic significance must

be reviewed with management. In addition to top management, production management should also be consulted. They are the people who will have to make the plan work. In some situations, it may also be beneficial to obtain the objective appraisal of an external or internal consultant.

9.5 ROUGH-CUT MANUFACTURING PLANNING

There are some methods for shortcutting the rigors involved in generating a detailed manufacturing plan. These methods primarily involve a crude projection of ratios and trends. In its crudest form, a rough-cut plan may simply be based on a linear calculation. This linear approach would assume that to double production, one must double the manpower, equipment, and facilities.

A little more sophistication can be added to this method by utilizing past history as well as the present. Output per person or output per square foot can be calculated for each of the past few years. Output could be defined as either dollars or product units. Awareness of trends regarding these data may enable more accurate rough-cut planning.

Rough-cut planning can be related to the accuracy versus timeliness conflict. Rough-cut planning requires a minimal amount of data and a minimal number of calculations. In turn, the time required to generate a plan is also minimized. However, there is a trade-off. Minimizing the effort and time will also minimize the accuracy.

The rough-cut planning approach is limited. It is difficult to calculate the impact of new products, new technologies, new techniques, or new production strategies. A change in the construction of one product or in one facet of the production operation cannot be integrated into a rough-cut plan. This ratio-trend approach is simply a projection of the old operation into the future. The impact of product and process changes can only be generalized.

The ratio-trend approach may be a necessity if speed is of the essence. At least this approach will provide a ballpark estimate with which to work. Such an estimate may be particularly useful in an environment unencumbered by significant product and process changes.

Rough-Cut Manufacturing Planning 147

In many ways, generating a rough-cut plan is the same as generating a detailed plan. Time must be spent in collecting the data, calculating requirements, and determining strategy. The major difference is that minimal detail requires minimal time.

The ratio-trend approach was succinctly illustrated by Richard Muther in his book *Systematic Layout Planning* (see Table 9.1). His example illustrates the ease with which a rough-cut plan can be developed. This rough-cut approach consists of three forms of data: (1) base data, (2) derived data, and (3) forecasted data. The base data are the foundation of a rough-cut plan. These facts and figures describe current and past operating conditions. In this case, the base data consist of sales dollars, units produced, number of employees, and space utilized.

From these facts, new information can be derived. As in every plan, it is necessary to establish a relationship between manufacturing requirements and marekting projections. This relationship is expressed as a ratio such as net sales per employee or net sales per square foot.

Forecasted data take two forms: a marketing forecast and an operations forecast. Typically, a rough-cut approach utilizes gross generic product groupings. The marketing forecast is relatively straightforward: a projection of sales dollars or units to be produced for each generic group. However, it is very important that the base data and the marketing forecast be consistent. Both must include or exclude the same specific products, or product models, within their definition of a generic product group.

In addition to the marketing forecast, the operating ratios also must be forecasted. Unfortunately, there will not usually be enough time or data available in order to do any sophisticated trend analysis. Most likely the current operating ratios will have to be reviewed by management. Management can then establish goals and dictate the operating ratios to be used for long-range planning. The operations forecast and the marketing forecast can then be combined to develop requirements. In this case, head count and space requirements are projected. Notice the three different methods for determining space.

Even with a relatively simple rough-cut plan, there are some problem areas. When determining current and past floor space, one should consider the utilization of that space. Assume a factory's capacity is double what it is producing, i.e., 50% utilization. In this situation, the derived data (sales dollars

Table 9.1 Rough-Cut Plan by Ratio Trend Approach

	4 years ago	2 years ago	This year	Forecast 2 years from now	Forecast 5 years from now
Base data[a]					
Net sales ($)	385 M	855 M	1,300 M	1,800 M[b]	2,600 M[b]
No. of pieces (units) produced	805	1,720	2,660	3,800[b]	5,500[b]
No. of shop employees	15	35	51	(1 ÷ 7) 67[c]	(1 ÷ 7) 93[c]
No. of office employees	8	11	16	(1 ÷ 8) 67[c]	(1 ÷ 8) 24[c]
Square feet in service shop	5,250	11,000	17,000	(1 ÷ 9) 22,500[d] (3 × 11) 23,400[d] (2 × 13) 24,700[d]	(1 ÷ 9) 30,600[d] (3 × 11) 32,600[d] (2 × 13) 35,800[d]
Square feet in office	750	1,600	2,100	(1 ÷ 10) 2,840[d] (4 × 12) 2,700[d]	(1 ÷ 10) 3,900[d] (4 × 12) 3,200[d]
Derived data[e]					
Net sales ($) per employee (shop)	25,600	24,400	25,500	27,000[f]	28,000[f]

Net sales ($) per employee (office)	48,200	77,600	81,300	90,000[f]	110,000[f]
Net sales ($) per square foot (shop)	73	78	77	80[f]	85[f]
Net sales ($) per square foot (office)	514	535	620	635[f]	665[f]
Square foot shop per shop employee	350	315	334	350[f]	350[f]
Square foot office per office employee	94	145	131	135[f]	135[f]
Square foot shop per piece (unit) produced	6.5	6.4	6.4	6.5[f]	6.5[f]

[a] Taken from company records or estimates of officials.
[b] Forecast sales and production.
[c] Manpower calculations based on forecasts of footnote b and ratios of footnote f.
[d] Square foot calculations based on figures of footnotes b and c and ratios of footnote f.
[e] Ratios derived from figures in base data.
[f] Ratios projected—based on trends and anticipated products and methods.

Source: From R. Muther, *Systematic Layout Planning*, Van Nostrand Reinhold, New York, 1961 & 1973.

per square foot, square foot per unit produced, square foot per employee) is misleading unless utilization is taken into account. The effect of ignoring utilization in this case would lead to projecting double the necessary square footage. Perhaps this is an extreme case, but it does demonstrate the need to consider production capacity versus actual production when one is developing operating ratios for planning future requirements.

10
Generating a Production Schedule

10.1 DETERMINING A TIME FRAME

Production managers rarely discuss capacity in terms of units per year. It is more likely that production rates are described in units per shift, units per day, units per week, or perhaps even units per month. Unfortunately, marketing forecasts are usually presented in terms of units per year or units per quarter. Therefore, it is usually necessary to reexpress marketing forecasts in terms of the time frame best understood by manufacturing management.

There is usually a generally accepted time frame when throughput capacity is discussed. This time frame is usually expressed by a department manager when he or she is asked, "What is your department's capacity?" Most likely the response will express an output per a given time frame.

The time frame utilized when discussing capacity is probably the same as that used for other reporting systems. The production schedule is of particular significance. In effect, it expresses the factory's goals. If the schedule is in terms of units per day, then production management will think in terms of units per day.

Production reports, equipment reports, and labor reports are also an indication of manufacturing's time frame. It must be noted, however, that the frequency of the report does not usually indicate time frame, but rather the content of the report. For instance, even a monthly report may indicate average daily production rate.

A manufacturing plan should be developed utilizing a recognized manufacturing time frame. Although the time frame will not alter requirements (head count, equipment, and space), it should expedite understanding and acceptance of those requirements. The plan is more readily audited when the forecasts, the equipment-hours, and the man-hours are all expressed in the context of a familiar time frame.

The manufacturing plan can be determined within a shift, day, week, or month time frame. The same logic and calculations apply no matter which time frame is chosen. The objective is to describe tomorrow's requirements utilizing today's terminology. However, there are some advantages to using an average weekly time frame in calculating manufacturing requirements. Mainly, these advantages relate to the ease with which the calculations can be made, the data interpreted, and the plan revised. For instance, hours available per week are recognized and manipulated easily. On a two-shift operation, there are 40 hours available per worker and 80 hours available per machine.

The ability to manipulate overtime is also simplified. Within a weekly time frame, weekend overtime can be easily added to the hours available. It is not as easily calculated or interpreted in a plan done for an average day rather than an average week.

Planning for an average week is also easier than planning for quarterly or annual requirements. The standard hours required per quarter or per year become too large to manipulate. The numbers could exceed the size alloted on a standard spread sheet. Even the hours available become unwieldy, roughly 2000 hours per person per year or 4000 hours per equipment per year on a two-shift basis.

This text will present a format for generating manufacturing requirements based on a typical week within the forecast period. With minor and probably obvious modifications, the same format can be utilized for any time frame.

10.2 CONVERTING THE PRODUCT FORECAST INTO A FACTORY-BUILT PLAN

10.2.1 Time Frame

Once a time frame has been established for planning purposes, the product forecast should be in terms of that time frame. For instance, a marketing forecast expressed in product units per year must be converted to average product units per week. Converting a forecast of units per year or units per quarter into units per week is relatively straightforward. An average weekly production rate is calculated by dividing the units per year by the weeks per year, although it must be recognized that for manufacturing there are not 52 weeks available per year. Ten holidays and a 1-week plant shutdown will leave only 49 weeks available for production. The production time available should not include holidays or plant shutdowns.

In large manufacturing concerns, some portion of the manufacturing operation may have an annual shutdown, while others do not. When a manufacturing plan is being generated, this partial shutdown must be taken into account. Two approaches can be taken. One is to calculate a separate average weekly product forecast for the operation with the shutdown. The other approach is to account for the shutdown by adjusting the utilization and uptime factors.

Table 10.1 illustrates the calculations required to determine weeks available by quarter and by year. Note that the weeks available range from 11.6 to 12.6 weeks per quarter. Obviously there is a direct correlation between the weeks available and the output capacity. A factory cannot produce as much in 11.6 weeks as it can in 12.6 weeks.

Usually the marketing forecast does not take into account the availability of manufacturing time. The marketing forecast is usually a projection of the product quantities to be shipped within a given time period. Except for perishable goods, it is usually not necessary to manufacture the product during the stated time period. The production rate (and hopefully growth) should be even throughout the four quarters. It may be necessary for the planner to adjust the marketing forecast to take into account the production weeks available.

Table 10.2 demonstrates the impact of weeks per quarter on a production forecast. Section 1 is a quarterly forecast for

Table 10.1 Determining Productive Weeks Available per Quarter and per Year

Quarter 1

    1 day — New Year's Day
    1 day — Washington's Birthday

    2 holidays

Quarter 2

    1 day — State Holiday
    1 day — Memorial Day

    2 holidays

$$\frac{2 \text{ holidays}}{5 \text{ days/week}} = 0.4 \text{ weeks unavailable}$$

    13.0 weeks/quarter
   − 0.4 weeks unavailable

    12.6 weeks available

Quarter 3

    1 day — July 4th
    1 day — Labor Day
    1 week — Summer shutdown

    1.4 weeks unavailable

    11.6 weeks available

Quarter 4

    2 days — Thanksgiving
    2 days — Christmas

    0.8 weeks unavailable

    12.2 weeks available

Weeks/year

    12.6 weeks—quarter 1
    12.6 weeks—quarter 2
    11.6 weeks—quarter 3
    12.2 weeks—quarter 4

    49.0 weeks/year

Table 10.2 Impact of Weeks per Quarter on a Production Forecast

		Quarter				Total year
		1	2	3	4	
Section 1						
Product forecast	Model A	950	1000	1050	1100	4100
	Model B	850	1000	1100	1200	4150
	Model C	1260	1260	1260	1260	5040
Section 2 (no. of weeks/quarter)		12.6	12.6	11.6	12.2	49
Average weekly rates	Model A	75.4	79.4	90.5	90.1	
	Model B	67.5	79.4	94.8	98.4	
	Model C	100	100	108.6	103.3	
Section 3 (smoothing factor)		1.029	1.029	0.947	0.996	
Smoothed weekly rates	Model A	77.6	81.7	85.7	89.7	
	Model B	69.4	81.7	89.8	98.0	
	Model C	102.9	102.9	102.8	102.9	

three product models: product A, B, and C. Products A and B show smooth growth trends per quarter. Product C remains flat. In section 2, the product forecast is expressed in units per week. It is calculated by dividing the units per quarter by the weeks available per quarter. Notice the impact on model C's production rate. The weekly production rate shows a 8.6% increase in the third quarter yet the marketing forecast is flat. The same surge effect occurs on the other products, although it is somewhat obscured by their growth rate.

Most manufacturers will build in advance of holidays or shutdowns. Short-range production scheduling takes these times into account; so should a medium- to long-range plan. The weekly production rates can be smoothed to reflect a more realistic production schedule. For instance, product model C could be produced at a rate of 102.9 per week throughout the year (5040 per quarter divided by 49 weeks per quarter equals 102.9). Product models A and B must also be smoothed by the same percentage. This percentage can be calculated by multiplying the weeks per given quarter (four quarters per year) and then dividing by the weeks available per year. The resulting ratio describes a relationship between the quarter in question and an average quarter.

Another way to look at this relationship is to compare the actual quarters to an average quarter. First calculate the average weeks per quarter by dividing the weeks per year by the number of quarters (in this case, 49 divided by 4 equals 12.25 average weeks per quarter). The smoothing factor then becomes the actual weeks per quarter divided by the annualized average weeks per quarter (in this case 12.25).

Section 3 of Table 10.2 demonstrates the effect that the smoothing factors have on the weekly rates. Not only is product C more equally distributed throughout the year, but so are the growth rates for products A and B.

The planner should also recognize that the production rate for a typical week, smoothed or not, is a midpoint forecast for the period being studied. In some situations, a midpoint forecast is insufficient for manufacturing planning. For instance, the average week for an annual forecast would most likely reflect the typical production rate around late June or early July. However, the goal for the manufacturing plan may be to project manufacturing requirements at the end of the year, not the middle.

To determine a typical week's production at the end of the year, product trends must be evaluated. Usually a year end

Converting the Product Forecast

weekly production rate may be extrapolated by projecting these trends on a graph. Assume that a 2-year marketing forecast must be converted to manufacturing requirements. The forecast presents the first year production volumes by quarter and the second year forecast contains only the production requirements for the year. Table 10.3 contains a year 2 annual forecast that is in addition to the quarterly forecast for the one year presented in Table 10.2.

Fig. 10.1 demonstrates a significant difference between the midyear production rate and the end-of-year production rate. The solid dots represent weekly production rates that were determined by calculations. Points denoted by an x are the derived end-of-year weekly production rates. Since midyear, product A has shown a significant increase, whereas product B has shown a marked decline. A situation such as this could have a radical impact on the manufacturing requirements. Consideration must be given to trends, especially when planning is based on annual rather than quarterly forecasts.

10.2.2 Components Forecasting

The basic components forecast relies on two pieces of information. First, it is necessary to have a product forecast—in this case, an average weekly production rate for each time period. Second, a listing of the components or subassemblies per product must be available. As discussed in Chapter 5, it is necessary to convert the company's product requirements into the factory's product requirements.

Table 10.3 Year 2 Annual Forecast

	Model	Year 2 total
Product forecast	A	4900
	B	4650
	C	5040
Average weekly rates	A	100
	B	94.9
	C	102.9

Generating a Production Schedule

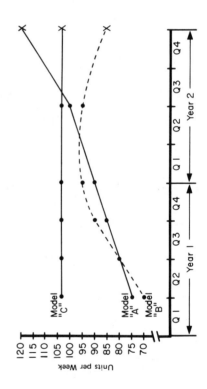

Converting the Product Forecast

Weekly rates	Year 1				Year 2	
	Q1	Q2	Q3	Q4	Q2	Q4
Model A	75	80	85	90	100	120
Model B	70	80	90	95	95	85
Model C	102.9	102.9	102.9	102.9	102.9	102.9

Quarterly rates	Year 1				Total year 2
	Q1	Q2	Q3	Q4	
Model A	945	1008	986	1098	4900
Model B	882	1008	1044	1159	4655
Model C	1296	1296	1193	1255	4165

Fig. 10.1 Weekly production rates by quarter.

The process of developing a subassembly forecast can range from extremely simple to extremely complex. It is entirely dependent upon the nature of the product, its intricacies, and the extent of in-house component/subassembly manufacturing. Constructing a component/subassembly forecast can be simplified by proper organization. By using a calculator and a spread sheet, even a cumbersome exercise is reduced to a straightforward procedure.

A master sheet should be drawn up that illustrates the relationship between the component level and the end product. A separate spread sheet should be constructed for each major manufacturing department. Separate component/subassembly forecasts are especially needed when manufacturing is horizontally integrated, i.e., when dedicated subassembly plants supply multiple final assembly plants. Each dedicated subassembly plant will require its own manufacturing plan. Therefore, when converting a product forecast to a component forecast, one should be concerned with only the components produced by the manufacturing plan under study.

The master sheet should list the components, whether part numbers or descriptions, across the top of the matrix. The end products should be listed along the side of the matrix. The ratio of component per product should then be entered in the intersecting block (see Table 10.4). In this example, product A gets one each of components 1 and 2. Product B also gets one each of components 1 and 2. In addition, one-half of the product Bs that are shipped contain an additional component 2 and 3. Product C is typically shipped with one component 1 and one component 3 and half of the time with component 2 as well.

A component forecast can be calculated by setting up a similar matrix on a separate sheet. The main difference is that

Table 10.4   Components per Product

Product	Component		
	1	2	3
A	1	1	0
B	1	1.5	0.5
C	1	0.5	1

Converting the Product Forecast

Table 10.5 Year 2 Fourth Quarter Average Weekly Requirements

Product	Forecast (quantity/week)	Component		
		1	2	3
A	120	120	120	0
B	85	85	127.5	42.5
C	102.9	102.9	51.5	102.9
Subtotal (component quantity/week)		307.9	299	145.4

the first column should be available to enter the product forecast quantities. By multiplying the forecast quantities by the ratios noted in the master sheet, one can establish the components required for each product. Once the components are established for each product, they can be summed into the total components required for all products.

Table 10.5 demonstrates the conversion from products requirements to components requirements. Although a separate sheet must be done for each time period, this example utilizes the year 2 fourth quarter forecast as developed in Fig. 10.1. The decimals have been carried out to one place, which is probably not necessary. However, it may be critical in situations where component quantities are relatively small.

10.2.3 Factoring the Component Forecast

The components forecasts must be adjusted to reflect what the factory will actually build. At this point, the component forecast is merely a reflection of the component quantities necessary to support product shipments. Most likely, this forecast will have to be adjusted to reflect: (1) additions such as spare parts, expansions, and second-source items, and (2) deletions such as offloads to vendors or other factories. All of these considerations fall into these two categories: they are either additions or deletions to the component level forecast. These additions or deletions may be quantified either as an actual quantity per week or as a percentage of the components required to support product shipments. For instance, in a division of an automobile company, it may be necessary to produce more motors than are needed to produce the company's cars. The factory that produces the motors may have to build 5% more motors for spares and an additional 100 per week for another division. Similarly, policy may also dictate the percentage of motors to be built by outside vendors or the actual weekly rate to be built by another factory.

It is possible that one component could have both actual and percentage additions as well as deletions by percentage and actuals. When a component requires multiple factors, the planner must be careful to apply those factors in the proper sequence. For instance, a spares percentage would probably be calculated on top of total requirements, including those built for other divisions and those offloaded for manufacture by vendors.

The impact of additions and deletions to a component forecast is illustrated in Table 10.6. It is an extension of the same spread sheet that was constructed in Table 10.5, although typically all of these calculations would take place on the same spread sheet.

Table 10.6 Additions/Deletions to the Components Forecast

	Component		
	1	2	3
Second-source additions (actual)	+52	+21	
Spares addition (percentage)			+10%
Offload (actual)		−100	
Offload (percentage)	−50%		
Adjusted total (component quantity/week)	180	220	160

Converting the Product Forecast   163

10.2.4 Component Lead Time

Consideration should be given to relabeling the component forecast that has been developed. In many cases, the time frame for the component forecast should not be the same as that for the product forecast. The component forecast should be relabeled to reflect the lead time between producing the component and shipping it with a product.

With any product, it is necessary to build the components before one can produce the end product. In some industries, there is a significant delay between when the subassembly is produced and when it is shipped with the product. The component/subassembly lead time is particularly significant when this lead time becomes close to the time interval being forecasted. A 13-week lead time in a quarterly forecast is a good example. The components must be manufactured in the third quarter in order to be shipped with the product in the fourth quarter. Fig. 10.2 demonstrates the effect of component lead time on a

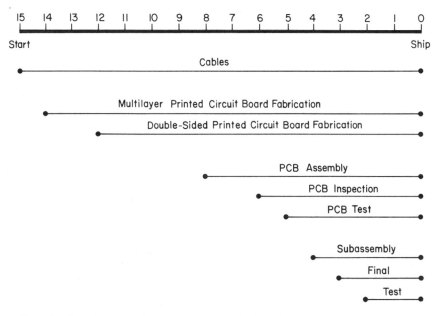

Fig. 10.2 Effect of component lead time in weeks prior to customer ship.

complex electronics product. The wiring, cables, and harness assemblies have a 15-week lead time. The raw circuit boards average 13 weeks—a full quarter from the start of fabrication until they are shipped. Although only 4 to 6 weeks of the time is spent in fabrication, the final 8 weeks is spent in inserting the components, testing the board, and inserting the board into final and subassemblies.

For the sake of demonstration, assume that components 1, 2, and 3 of the earlier examples are produced 10 weeks in advance of product shipments. Therefore, the majority of fourth quarter product shipments will consist of components made in the third quarter. The components forecast should then be relabeled "year 2 third quarter" to account for this lead time. The major objective of a manufacturing plan is to ensure that assets are available on time.

11
Generating Manufacturing Requirements

11.1 CAPACITY REQUIREMENTS BY OPERATION

A comprehensive and detailed manufacturing plan must delineate the requirements of each individual operation. Although there are rough-cut methods for estimating manpower and space requirements by work center or department, accurate projections should be summarized from the requirements of each individual operation. Equipment requirements can be calculated only at the individual operation level.

11.1.1 Calculation of Standard Labor Hours and Equipment Hours

As discussed earlier, the essence of a manufacturing plan is the relationship between the product and the process. Although sometimes complex, this relationship can best be depicted and manipulated when formatted as a matrix on a standard spread sheet.

It is important that the spread sheet for the planning matrix be properly formatted. The columns should be labeled to represent the products or components to be processed. The rows should be labeled for each step in the production process. These steps should be the same as those defined in the process flow chart with one exception. In most cases it will not be necessary to detail each transportation step on the planning matrix. Handling time is included in the direct labor utilization factor. Handling space is usually included in the footprint required of each operation. The handling equipment itself is dependent on the number of operations it must service. Therefore, the planning matrix should contain only major processing steps, such as operations, inspections, major delays, and storage areas.

For some processing steps one must be careful to distinguish between labor content and equipment content. This situation occurs when there is not a one-to-one relationship between machine and operator within each operation. In a setting where this situation is infrequent, its occurrence can be handled by separate entries on the matrix. Although both line entries will contain the same operation description, one should be clearly labeled "labor," while the other clearly labeled "equipment."

This double entry procedure may prove unwieldy for the capital-intensive or process-type industry. In these industries it is not unusual for there to be a big difference in the labor and equipment content in the majority of operations. In such an environment, equipment planning and labor planning should be kept separate. Although the items to be processed and the processing operations will be the same, one matrix will contain equipment hours per item processed, while the other matrix contains labor hours per item processed.

In addition to the major processing steps, points of significant yield loss should be entered. For instance, an inspection station may reject and scrap a significant portion of the items that have been processed until that point. In this case the word "yield" should be entered on the row following the inspection operation. This will force an entry to be made as to the percentage yielded for each component at that operation.

A similar situation occurs with rework. Prior to the operations affected by rework, the term "rework" should be noted on the planning matrix. It should also be noted when the reworked items affect more than one operation. A simple line, arrow, or checkmark should be sufficient to note the affected operations.

Capacity Requirements by Operation

The first row of the planning matrix should be left blank. This row will be used to enter the start quantity of each component to be processed.

The planning matrix format is exhibited in Table 11.1. The operations are taken from the flow chart that was presented in Chapter 6 (Table 6.2).

Once the rows and columns on the matrix have been labeled, a number of duplicates of that matrix should be made. A copy should be made for each time period to be studied. There should be some extra copies to take into account errors and the inevitable future recalculations.

One copy of the matrix should be titled master planning matrix. As a master sheet this matrix will contain the base data needed for planning each time period. The master matrix will contain the processing hours for each item or component at each step of the manufacturing process. This matrix primarily requires reformatting the processing data collected earlier. The matrix is not much more than an organized presentation of industrial engineering's direct labor standard hours and reasonable expectancy times, or equipment cycle times.

An example of the master planning matrix or input matrix can be seen in Table 11.2. Note that the line entries for rework and yield percentages remain blank. Yield and rework percentages are variables and may vary with each time frame that is forecasted.

Table 11.1 Planning Matrix Format

	Component		
Start quantity	A	B	C
Operation #1			
*Rework (%)			
Operation #2			
Operation #3			
Inspection #1			
*Yield (%)			
Operation #3			

Table 11.2 Example of Master Planning Matrix

Start quantity	Component			Total standard hours
	A	B	C	
Operation #1 *Rework (%)	0.015	0.02	0.065	
Operation #2	0.25	0.095	0.355	
Operation #3	0.009	0.009	0.015	
Inspection #1 *Yield (%)	0.125	0.125	0.125	
Operation #3	0.075	0.05	0.045	

The actual exercise of calculating manufacturing requirements should be done on a projection matrix. A copy of the formatted but blank planning matrix should be titled for the first forecast period being studied. It should reflect the forecasted production period for the factory, not necessarily the shipments period for the end products.

Utilizing this matrix approach and calculating requirements are part of a step-by-step process. Completed in sequence, these steps should simplify an otherwise complex process.

Step 1. Insert Yield Data

For every row labeled "yield," an entry should be made for each item processed. The entry should reflect the percentage of the item's production quantity which will pass on through to the next operations. Like the forecast itself, the yield figures may represent a goal.

Once the yield rates have been entered throughout the process, the overall yield rate for each item processed must be calculated. For any item produced, the overall yield rate is a multiple of the individual yield rates encountered throughout the process. For instance, assume a production item was inspected three times throughout the manufacturing process. At all three inspections, 90% of the items passed while 10% was scrapped. In effect, the overall yield for that item would be 72% ($0.9 \times 0.9 \times 0.9 = 0.72$).

Capacity Requirements by Operation

Step 2. Calculate Start Quantities

As demonstrated, the yield rates encountered during the production process can radically reduce the effective production rate of that process. In order to achieve the desired output, the quantities that are input at the beginning of the production process must take into account these inevitable losses. The forecasted production item or component quantities must be divided by the overall yield ratio. In the previous example, if one wished to produce 100 units, roughly 139 units would have to be started ($100/0.72 = 139$).

The start quantity for each item in the forecast should be calculated. Once calculated, the start quantities should be entered in the top row of the projection matrix.

Step 3. Calculate Labor and Equipment Hours

The total equipment hours or labor hours required of an operation are simply the sum of the hours required to process the forecasted quantity of each product/component item through the operation. Therefore, the first priority is to determine the hours required to process the forecasted quantity of each item at each operation. The start quantity in the projection matrix should be multiplied by the hours per item on the master matrix. The result should then be entered on the projection matrix.

When a yield percentage is encountered, the production quantity should be multiplied by that yield factor. The new yielded production quantity can then be entered next to the yield percentage on the projection matrix. For the rest of the plan or until the next yield rate is encountered, the yielded quantities will be used for determining labor and equipment hours.

Once the projection matrix has been filled in, the total hours required of each operation can be calculated. On the projection matrix, the last column should be labeled "total." The hours within each row can then be summed across and the total weekly standard hours can be entered in the total column.

Continuing the earlier example, the completed projection matrix is illustrated in Table 11.3.

11.1.2 Labor Requirements

In order to develop the projected head count, it may be useful to construct a labor matrix. This labor matrix would simply be

Table 11.3 Completed Projection Matrix

	Component			Component			Total standard hours
	A	B	C	A	B	C	
Start quantity	250	100	200				
Operation #1	0.015	0.02	0.065	3.75	2	13	18.75
*Rework (%)	10	10	10				
Operation #2	0.25	0.095	0.355	68.75	10.45	78.1	157.3
Operation #3	0.009	0.009	0.015	2.475	0.99	3.3	6.765
Inspection #1	0.125	0.125	0.125	34.375	13.75	27.5	75.625
*Yield (%)	90	95	90				
Operation #3	0.075	0.05	0.045	16.875	4.75	8.1	29.725
Output quantity	225	95	180				

Capacity Requirements by Operation

an extension of the projection matrix. While the projection matrix contains the direct labor standard hours required, the labor matrix will be used to calculate the exact number of employees required.

The labor matrix will utilize the total weekly standard hours of each operation as developed in the projection matrix. The weekly standard hours will be factored for performance, utilization, attendance, and potential productivity improvements. The resulting hours per operation can then be summed to reveal the total weekly factored hours required for the entire work center or department. The man-hours per week required can then be divided by 40 hours/week per person to determine the head count for the work center or department.

The labor matrix and its relationship to the projection matrix are exhibited in Table 11.4. Note that the operations on the labor matrix are labeled on the same rows as on the projection matrix. In particular, note that operation #2 contains only one entry on the labor matrix. Through careful relabeling of the labor matrix, one can avoid inadvertently carrying equipment hours into the labor calculations.

The labor matrix is designed so that every block can be used which may not be necessary. For instance, assume that performance is the only factor to vary by operation. The other factors of utilization, attendance, and productivity improvement are known or projected for the department as a whole but not at the individual operation level. In this situation it will be simpler to factor the operation hours by performance, sum them, and then factor this total for utilization, attendance, and productivity. The number crunching effort could be significantly reduced.

Care must be taken when calculating factored hours from standard hours. Standard hours must be divided by performance, the result divided by utilization, and that result divided by the attendance factor, not necessarily in that order. The hours factored for performance, utilization, and attendance must be multiplied by the productivity improvement factor. The productivity improvement factor will reduce the hours required, while the other factors will increase the hours required.

In most environments the hours available per worker per week will be 40. Even when two shifts are being run, any one worker will put in only 40 hours/week. Most likely this 40-hour figure will apply across all operations. Although, in some cases, there may be planned overtime at specific operations, usually planned overtime should not be included in the manufacturing

Table 11.4 Relationship of Labor Matrix to Projection Matrix

	Component			Component			
	A	B	C	A	B	C	Total
Start quantity	250	100	200				
Operation #1 *Rework (%)	0.015 10	0.02 10	0.065 10	3.75	2	13	18
Operation #2 (equipment)	0.25	0.095	0.355	68.75	10.45	78.1	157.3
Operation #2 (labor)	0.009	0.009	0.015	2.475	0.99	3.3	6.765
Inspection #1 *Yield (%)	0.125 90	0.125 95	0.125 90	34.375	13.75	27.5	75.625
Operation #3	0.075	0.05	0.045	16.875	4.75	8.1	29.725
Output quantity	225	95	180				

Capacity Requirements by Operation

Start quantity	Perform-ance	Utili-zation	Attend-ance	Productivity improvement	Required hours	Available hours	Head count
Operation #1 *Rework (%)	90	80	92	90	25.48	40	0.6
Operation #2 (equipment)							
Operation #2 (labor)	95	75	92	95	9.80	40	2.9
Inspection #1 *Yield (%)	90	75	92	95	115.69	40	2.9
Operation #3	90	70	92	95	48.72	40	1.2
Output quantity							

[a] Department head count = 5.

plan. Whenever possible, future operations should not be based on current inefficiencies.

Manpower requirements are also effected by worker flexibility or mobility within the factory. At some point a fractional head count must be rounded upwards to the next whole number. Where workers are flexible within a work center and capable of interchanging jobs, then the factored hours can be summed for all operations throughout the work center. However, in other situations, specific operations may require a dedicated and specialized operator. These operators may not be transferable to any other task. Therefore, any fractional requirements of these workers must be rounded up to the next whole number at the operation level. (See operation 2 in the labor matrix example.)

The same flexibility or lack of flexibility can occur between work centers and departments. Where the work centers are flexible throughout a plant, the work center or departmental head count could be calculated to one decimal. Only the total plant head count would have to be rounded to the next whole integer.

11.1.3 Equipment and Space Requirements

Equipment and space requirements can be handled in much the same fashion as labor requirements. The projection matrix which, for this purpose, contains the weekly equipment hours required per operation will be interfaced with an equipment matrix. The equipment matrix will be used to convert hours to equipment quantities and, in turn, space requirements.

The equipment matrix should be labeled so as to line up properly with the projection matrix. The operations should be listed on the same rows as they are listed on the projection matrix. In Table 11.5, note that the equipment matrix is concerned only with equipment hours from the projection matrix. The labor hours in operation 2 have not been mistakenly carried over onto the equipment matrix.

Similar to the labor matrix, the columns on the equipment matrix must be labeled for the productivity factors (in this case, performance, uptime, and productivity improvement). These factors will be used to convert equipment standard hours into equipment factored hours. The standard equipment hours for each operation must be divided by the performance factor and

Capacity Requirements by Operation

the result divided by the uptime factor. The resulting hours are then multiplied by the productivity improvement factor to determine the total factored weekly hours per operation.

To determine the number of pieces of equipment required, it is first necessary to determine the hours available per piece. The hours available per week per piece of equipment is most dependent upon the number of shifts. A one-shift operation will typically yield 40 hours/week available production time. By running a second shift, each piece of equipment becomes operational for 80 hours/week. However, in many cases, a third shift will not yield 120 hours/week available. Lunch breaks may cut into available time, assuming that the equipment is operator dependent.

In some cases, a first shift runs from 7:00 to 3:30, and a second shift from 3:30 to 12:00. Both contain $8\frac{1}{2}$ hours. Therefore, a half-hour lunch does not detract from the 40- or 80-hour week. However, the addition of a third shift will contribute only 6.5 additional available hours per day for a total of 22.5 hours or 112.5 hours/week. Even if each shift is equalized at 8 hours each (7 to 3, 3 to 11, 11 to 7), a half-hour lunch on each shift leaves 7.5 hours available.

Most of the time, an entire work center or department will work the same number of shifts. However, on occasion it is necessary to run a major piece of capital equipment for multiple shifts, while the rest of the department remains on a single shift. On the equipment matrix, hours available can be handled independently for each operation. The number of pieces of equipment required can then be calculated by dividing the total factored hours per piece by the hours available. As for calculating manpower, it is not necessary to calculate the equipment required beyond one decimal point. Most of the time fractional equipment requirements will have to be rounded to the next highest whole number. Of course, there are exceptions. Sometimes it may make more sense to run additional overtime rather than to purchase a new piece of equipment.

Having established equipment and work bench requirements, it is relatively simple to establish their space requirements. The number of pieces of equipment, times the space required per piece, will equal the total space required for that operation. The space required per piece is the footprint for that equipment. The footprint includes all of the space needed for the equipment itself, the operator, maintenance access, material handling, work-in-process, and operator egress.

Table 11.5 Equipment Matrix

	Component			Component		
	A	B	C	A	B	C
Start quantity	250	100	200			
Operation #1	0.015	0.02	0.065	3.75	2	13
*Rework (%)	10	10	10			
Operation #2 (equipment)	0.25	0.095	0.355	68.75	10.45	78.1
Operation #2 (labor)	0.009	0.009	0.015	2.475	0.99	3.3
Inspection #1	0.125	0.125	0.125	34.375	13.75	27.5
*Yield (%)	90	95	90			
Operation #3	0.075	0.05	0.045	16.875	4.75	8.1
Output quantity	225	95	180			

	Required hours	Available hours	Head count	On-hand
Operation #1 *Rework (%)	23.44	40	0.6	1
Operation #2 (equipment)	251.15	80	3.1	3
Operation #2 (labor)				
Inspection #1 *Yield (%)	106.44	40	2.7	4
Operation #3	44.82	40	1.1	2

[a]Square feet required for production area = 1775.

Capacity Requirements by Operation

Total standard hours	Performance	Uptime	Productivity improvement
18.75	90	80	90
157.3	85	70	95
6.765			
75.625	90	75	95
29.725	90	70	95

Required (%)	Equipment planned	Unit (square feet)	Space
59	1	175	175
105	4	200	800
67	4	150	600
56	2	100	200

The sum of the space required per each operation within a work center is the direct manufacturing space. It is a projection of the space required to support the direct production operations. However, it does not include the space required for indirect or support functions.

11.2 WORK CENTER REQUIREMENTS

A work center, or manufacturing department, usually consists of a grouping of all the operations needed to complete a specified portion of the product's construction. For example, in the assembly of circuit boards, there is usually a work center dedicated to automatic insertion operations and another work center for manual insertion operations.

The operations within a work center may only partially determine the assets that are required. In addition to the direct production operations, additional support space and equipment may be required.

When work center requirements are determined, one must be sure to include all supporting equipment that have not been included in the footprint at the operation level. Support equipment can be divided into two categories: equipment that will have no additional impact on space, and equipment that will have an impact on space.

When a manufacturing plan is done solely to determine space requirements, the planner should not be concerned with nonspace-consuming equipment. However, if the plan is primarily concerned with cost and acquisitions, the planner must be concerned with all major pieces of equipment.

Support equipment that takes up space include compressors, pumps, dust handlers, material handling equipment, and pollution control equipment. Under most circumstances, the space for these items should be taken into account at the operation level footprint. However, when a budget is being developed and procurement is being established, these items should not be overlooked. They are a necessary facet to the successful operation of the work center.

The space required for production and supporting equipment will make up most of the work center's space. The addition

Work Center Requirements

Table 11.6 Equipment and Space Requirements for Work Center

Labor	Equipment	Space
	Production equipment	
5	1—Operation #1	
	4—Operation #2	
	4—Inspection	
	2—Operation #3	
		1775
	Support equipment	
	1—Compressor	100
	1—Overhead conveyor	—
1	Supervisor's office	150
		2025

of a few specific areas will refine and complete a projection of square footage required for the work center.

Within each work center there will usually need to be a supervisor's office. A supervisor's office will usually contribute at least an additional 100 sq ft. The space required for a supervisor's office is dependent on a number of factors. In some industries, a supervisor's office may contain control equipment or computer equipment. In other industries, the supervisor's office may consist of a desk in a corner.

Within the work center, it may also be necessary to include space for break areas, a dumpster, lockers, miscellaneous supplies, and processing equipment storage. Specifically, processing equipment storage may be needed for molds, spray masks, nozzles, drill bits, blades, and other various interchangeable fixtures.

In the continuing example, Table 11.6 completes a work center's equipment and space requirements. Note that an overhead conveyor is an equipment requirement that has no impact on the square footage required.

11.3 TOTAL PRODUCTION LINE REQUIREMENTS

In most companies, a production line is synonymous with a manufacturing department. Made up of work centers, the production department is usually distinguished by the type of process and the type of product produced. Typically a manufacturing department or production line is distinguished by its own specialty. Many factories are made up of a number of production lines or manufacturing departments.

A computer manufacturer may have separate production departments for circuit board fabrication, circuit board assembly, and cables assembly. These departments are segregated primarily because of their process. The same computer manufacturer may have two separate final assembly production lines, one for each computer product. While the processing technique is essentially the same, the product distinguishes the two departments.

Any of these departments may contain a number of work centers. While the departments are usually run by a manager, a work center is usually the responsibility of a supervisor or foreman.

In order to determine the manufacturing requirements for a department, the planner must take into account more than just the requirements for each work center. Each production line and the entire factory must be supported by additional requirements not covered at the work center level. These additional requirements may be absolutely critical to the effective operation of the production line.

Whether discussing one production line or a collection of production lines as found in a large factory, the same factors apply. Whether these factors apply at the departmental level or the factory level will depend on the nature and management style of one's business. For instance, a tool crib may be decentralized and separately located within each manufacturing department. In another case, there may be one centralized tool crib to service the entire factory and numerous manufacturing departments.

The determination of production line or departmental requirements must be done first. The sum of the departmental requirements can then be factored to incorporate the support functions that are found at the factory level. Since these supporting functions can be found at either the factory or production level, they will be discussed only once. It is up to the planner to determine the appropriate point of inclusion.

Total Production Line Requirements   181

11.3.1 Labor

The bottom line head count for the production line or factory is made up of two components. One, which is by far the most significant, is the sum of the work center head counts. The second is a tabulation of miscellaneous labor, such as indirect and support functions or small operations that were not built into the manufacturing plan.

Some manufacturers build products that are no longer officially being sold. A major corporate customer may command enough clout to purchase an officially obsolete product. Although minimal, the labor required to support production of that dying product should be included in the manufacturing head count.

Similarly, the labor for any other off-line work should be included. Direct labor may be used for the off-line production of prototypes of new components or products. Some specialty work and customization may also be done off-line by direct labor. These somewhat random off-line operations would probably not be included in the detailed planning process and must therefore be added in.

Additional labor may be needed to support pilot production elsewhere. Although on the payroll of a production department, these employees may be on loan to another group. If the situation is expected to continue, these employees should be included in any projection of the department's head count.

Any function that falls under a production department manager's direct labor payroll may be included. The supporting labor such as that found in the tool shop, tool crib, or maintenance department should not be included in the direct production line requirements or even the direct factory requirements. These indirect employees will be discussed in Section 11.4.

Once a department's projected direct labor head count has been established, it should be put into perspective. The impact of the projected head count can be better visualized if it is compared to either the current head count or the projected head count from the previous time period. For instance, the projected head count for the fourth quarter of next year could be compared with the projected head count for the third quarter of next year. The growth rate (or decline) is easily reviewed and interpreted.

Table 11.7 illustrates an example of the departmental and factory labor projection. In the example, remember that the

Table 11.7 Example of Departmental and Factory Labor Projections

		Year 1				Year 2			
Department	Current	Quarter 1	Quarter 2	Quarter 3	Quarter 4	Quarter 1	Quarter 2	Quarter 3	Quarter 4
Fabrication	75	80	80	82	85			90	92
Subassembly	51	55	58	60	62				67
Final assembly	60	60	65	68	70				77
Subtotal	186	195	203	210	217				236
Miscellaneous	14	15	17	20	23				29
Total	200	210	220	230	240				265

Total Production Line Requirements

hypothetical forecast contained quantities for the four quarters of year 1 and an annual quantity for year 2. The year 2 annual rates were calculated and trended to reflect a year 2 quarter 4 forecast. The year 2 manufacturing requirements have been developed only for the fourth quarter with one exception. The exception is the components that must be produced in the third quarter in order to support shipments in the fourth quarter.
In this situation the third quarter head counts have been trended to approximate the fourth quarter head counts. Without extrapolating the fourth quarter labor for that department, the total factory direct labor could not be tabulated for either quarter.

11.3.2 Equipment

Unlike labor and space, equipment requirements cannot be summarized into one neat number for each production line. Therefore, it is unlikely that the equipment requirements for all the production lines within a factory can be summarized on one page. Most likely a separate summary of equipment requirements will have to be done for each product line.
    A summary of production line equipment requirements should include the number of units of direct production equipment and directly related support equipment, such as dust collectors, paint shakers, and battery chargers. The listing should not include indirect support equipment such as the factory's boiler, transformers, and HVAC equipment. Service equipment that supports the factory as a whole will usually come under the jurisdiction of plant engineering.
    When a manufacturing plan requires a projection of plant service requirements, plant engineering should be consulted. By utilizing the manufacturing plan's projection of equipment, labor, and space, plant engineering can determine overall service requirements. The service requirements can then be included in a section of the documented manufacturing plan.
    The primary concern of a manufacturing plan is with direct manufacturing equipment. A cohesive summary sheet should be developed for each production line detailing each type of equipment, the quantity on hand, and the projected quantity required for each forecasted time period. To effectively illustrate trends and equipment utilization, it may be beneficial to present the equipment requirements as they were calculated to one decimal place.

A sample equipment report is presented in Table 11.8. Again, note that in year 2, the third quarter requirements were calculated, which accounted for production lead time. However, the bulk of the manufacturing plan is being done for year 2 quarter 4. Therefore, fourth quarter requirements can be extrapolated to ensure consistency.

For each production line, it may also be helpful to develop an "Additional Equipment Report." As an effective bottom line presentation, such a report would simply present when and how many pieces of additional equipment are required. The usefulness of such a report is illustrated in Table 11.9. Also notice that the year 2, first, second, and fourth quarter requirements have been extrapolated.

The presentation of an "additional equipment" list is effective because it is simple, although developing the report itself may not be so simple. The projected additional equipment listing is dependent on cost, critical impact, time, and management policy.

The acquisition of extremely expensive equipment may be delayed as long as possible. A major equipment expense can be postponed by maximizing the utilization of existing equipment. Production requirements could be met by overtime or additional shifts rather than by new equipment.

The addition of overtime or another shift should be noted on an additional requirements report as illustrated in Table 11.9. Of course, the planner must recognize that the addition of overtime or another shift is a strategic management decision. The presentation of strategy must be backed up by a presentation of requirements. The basic requirements are illustrated in Table 11.8, the sample equipment report.

The impact that a piece of equipment has on the production operation may also dictate acquisition policy. For some operations it is better to acquire equipment rather than to risk a bottleneck. Equipment cost may be negligible compared to the cost of missed production.

Equipment acquisition is a trade-off between cost and benefit. The purchase of costly equipment requires serious evaluation. Conversely, the purchase of an inexpensive work bench could be done at the slightest hint of an additional requirement.

In some environments, management policy will dictate at what point equipment should be acquired. For instance, in a high-growth environment, management may direct that it is time to acquire additional capacity whenever an operation hits 85%

Table 11.8 Sample Equipment Report

Equipment	Current	Year 1				Year 2			
		Quarter 1	Quarter 2	Quarter 3	Quarter 4	Quarter 1	Quarter 2	Quarter 3	Quarter 4
Mold	6	4.5	5.2	6	6.8			8.2	8.8
Drill	3	2.1	2.3	2.5	2.7			3.3	3.5
Assembly bench	12	10	10.5	11	11.5			13	13.5

Table 11.9 Additional Equipment Report

		Year 1				Year 2			
Equipment	Current	Quarter 1	Quarter 2	Quarter 3	Quarter 4	Quarter 1	Quarter 2	Quarter 3	Quarter 4
Mold	6			+1		+1		+1	
Drill	3						+OT	+OT	Second shift
Assembly bench	12						+1		+1

Total Production Line Requirements 187

of its capacity. Of course, there may be exceptions to the rule for specific pieces of equipment.

Lead time may also be a consideration in any presentation of equipment requirements. There may be times when the planner is interested in documenting when the equipment has to be ordered rather than when it has to be installed and operational. Essentially, such a report would look the same as the additional equipment report illustrated in Table 11.9 except for the time slot in which the requirement is recorded. For instance, a fourth quarter requirement may depend on a second quarter purchase. A detailed manufacturing plan must include a summary of equipment requirements. The specific format of that summary, and the information contained, will have to be tailored to a company's concerns and objectives.

11.3.3 Space

The space required to support a production line can be divided into four groups: (1) the sum of the space required for each work center—the direct production space; (2) major aisles; (3) employee services; and (4) plant services. In addition to these four areas, the total space requirement must include a factor for space utilization.

When the bottom line space requirements are calculated, the factoring for aisles, services, and utilization can be done at the production line level, at the department level, or at the factory level. Table 11.10 is a presentation of space requirements done for a factory. Notice that there are line items for each department or production line. Had the presentation been done for a production line rather than the entire factory, the space per work center would have been noted rather than space per department. The approach to take when summarizing space requirements can best be determined by the planner. Consideration should be given to the objectives of the plan and the audience for the information.

Either approach to summarizing space requirements requires that a subtotal be calculated for direct production space. This direct production space, whether a sum of work centers or a sum of departments, is the primary determinant of overall manufacturing space. The total departmental or work center space includes the footprint space for all production operations plus direct work center support space, such as a supervisor's office.

Table 11.10 Manufacturing Space Report

Department	Current	Year 1				Year 2			
		Quarter 1	Quarter 2	Quarter 3	Quarter 4	Quarter 1	Quarter 2	Quarter 3	Quarter 4
Fabrication	25000	26000	26000	27000	27000				30000
Subassembly	15000	15000	17000	18000	19000				24000
Final assembly	20000	22000	23000	24000	24500				27000
Subtotal	60000	63000	66000	69000	70500				81000
Major aisles	20000	21000	22000	23000	23500				27000
Plant services	10000	10500	11000	11500	11750				13500
Employee services	5000	5250	5500	5750	5875				6750
95% Utilization	5000	3150	3300	3450	3525				4050
Total	100000	102900	107800	112700	115150				132300

Total Production Line Requirements

Any projection of total manufacturing space requires an allowance for major aisles. These major aisles that have not been incorporated into the operational footprints are the plant's main thoroughfares. A major aisle usually services the plant as a whole, not just one department or work center.

Major aisle space can be factored as a ratio to the total direct production space. Based on the percentage of space occupied by aisles in the present facility, aisle space can be calculated for future facilities. In the example (Table 11.10), note that major aisles account for 20% of the current actual factory space. The direct production operations account for 60% of the factory's space. Therefore, the major aisle space can be calculated as one-third that of the production space. The direct production space times 0.33 will determine the major aisle space in this case.

Space for plant services can be calculated in much the same manner as for the space required for major aisles. Space for plant services can be calculated as a ratio of the direct production space.

When factoring for plant services, the planner will have to determine the level of detail desired. One factor can be used to account for all plant services or plant services can be factored by line item. Both approaches utilize the same technique. The plant service space is simply ratioed to the total factory space.

Under the jurisdiction of plant engineering, plant services include the tool room, the maintenance shops, boiler rooms, HVAC, guard shacks, battery chargers, vehicle storage, etc. Therefore, plant service space ratios should be developed in conjunction with the plant engineer to ensure validity.

In a detailed facilities plan, each of these plant services could be ratioed individually to the projected production space. However, in the example (Table 11.10) all of these plant services are summarized. In this case, the combined plant services account for 10% of current factory space. An additional one-sixth of the space, which is required for direct production operations, is needed for all of these services.

Employee services are different than aisle space and plant services. Employee services support people, not equipment or space. The space occupied by employee service functions is related directly to the number of people being serviced.

The employee service space requirement should be expressed as a square footage per employee. The space requirement per employee could be based on standards, government regulation, or accepted company practice.

Employee services could include washrooms, break areas, cafeterias, first aid rooms, time clock, personnel entrance, plus offices for plant and department managers. For any of these line items, the space per employee should reflect the capacity of the function, not necessarily the current situation. For instance, there are many situations where the factory is capable of handling many more employees than it currently houses.

Although current employee service space and percentages are calculated in Table 11.10, the employee service space projections are based upon a square foot per employee ratio and the head count from Table 11.7.

One more factor must be applied to take into account nonutilized space. Similar to aisles and plant services, nonutilized space can be calculated as a percentage of direct production space.

Although regrettable, nonutilized space is unavoidable. Footprints for each piece of production equipment takes into account the total space required to successfully operate that equipment. Unfortunately, in any plant layout, due to physical limitations, these footprints may not abut neatly. Inefficiencies in the utilization of space will occur due to the location of aisles, stairwells, doors, elevators, I-beams, etc. Also, the space occupied by some of these items such as stairwells and elevators should be included in the utilization figure if they have not been included elsewhere.

The example (Table 11.10) shows the current space utilization at 95%. Therefore, the nonutilized space accounts for 5% of the total factory space. Expressed differently, the nonutilized space occupies an additional 8.33% of the space occupied by the direct production operations.

11.4 ADDITIONAL REQUIREMENTS FOR EMPLOYEES AND SPACE

Typically, a manufacturing facility will house functions other than those directly related to manufacturing. For instance, every factory requires a warehousing operation of some sort. Also, it is not unusual for a manufacturing facility to house clerical, professional, and administrative personnel from other management functions such as finance, research and development, and marketing. It is highly likely that the facility will house

Additional Requirements for Employees and Space 191

manufacturing support groups such as manufacturing engineering, industrial engineering, production control, and cost accounting. These support functions cannot be ignored when an entire manufacturing facilities plan is being developed. However, in other cases, a manufacturing plan must cover only those assets directly related to the production of the product. Again the coverage of a manufacturing plan is determined by the needs of the organization and the purpose for which the plan was commissioned.

11.4.1 Clerical, Professional, Technical, and Administrative Functions

Although beyond the scope of direct manufacturing assets planning, office planning may be required. Estimates of offices (or laboratory) manpower and space needs can be generated through a number of techniques.

One approach is to utilize two ratios. One ratio is the square footage required per person for the function or functions being studied. Square feet per person probably will not change much from year to year in an office environment. The second ratio could be either an expression of revenue dollars per person or production units per person. In either case, the company's output either in revenue dollars or units of product is divided by the number of employees in the department being studied. Of course, this ratio technique assumes that there is a clear correlation between a department's staffing and the company's output.

Any derived relationship of output to people should take into account utilization. The ratio should be based on a department's capacity. For instance, a current engineering department may have staffing to meet a 10% growth in the company's revenue or production output. Therefore, any expression of output per person should be factored to take into account the department's full capability.

Output per person ratios can also be factored to take into account improvements in productivity. For instance, assume that there are 10 cost accountants in a $100 million company and it is now anticipated that computerized software will cut the manual efforts in half over 5 years. The output per person ratio can then be increased to take into account the improvement. If the company becomes a $200 million company, the number of cost accountants would remain the same.

A forecasted output divided by the output per person will yield the number of people required. The number of people required in a department must then be multiplied by the square feet per person to determine the space requirements. Admittedly, this approach is crude, however, a ballpark estimate may be sufficient for long-range planning.

To reduce inaccuracies, this ratio/trend approach should be tailored to each department being studied. For instance, manufacturing engineers should not be related to total revenue dollars. Manufacturing engineers should be related to in-house production build rates or production dollars. A company that opts for overseas manufacturing may be enhancing revenue while minimizing or eliminating the need for manufacturing engineers.

In some cases, it may be beneficial to relate clerical, administrative, professional, and technical functions to the direct manufacturing requirements. These functions should be expressed as a percentage of the direct manufacturing requirement. For instance, production engineering space may be a percentage of the manufacturing floor. Similarly, employee relations personnel may be expressed as a ratio of the direct labor head count.

When estimates of these peripheral functions are being generated, careful considerations should be given to the approach being used. One must consider the validity of the relationships being used. It must be recognized that different departments are driven by different factors. While plan engineering requirements may be directly related to factory space, the accounts payable function is more appropriately related to revenue dollars.

12
Warehouse Facility Planning

12.1 INTRODUCTION

The warehouse is a major component of many manufacturing facilities. Therefore, many manufacturing facilities plans would be incomplete without a detailed projection of warehousing requirements. Establishing a ratio of warehouse space to manufacturing space is usually inadequate. The same care and detail that go into the development of manufacturing requirements should also go into developing warehousing requirements.
    Warehouse facility planning is the process of determining future storage space requirements. Warehouse facility planning is done in two phases: data collection and requirements planning. Data collection involves gathering and formatting the base data. Properly organized, this base data will enable the generation of a warehousing requirements plan at any time. Requirements planning is the actual process of converting a marketing forecast into storage space requirements.

12.2 DATA COLLECTION

The following data are needed for warehouse space planning or for any storage space planning.

1. Material to be stored
2. Space required per item while stored
3. Length of time each item is to be stored

Developing these data may not be as simple as it first appears. Materials requirements change along with the product designs, the introduction of new products, and changes in the product mix. The length of time for storage is never a constant for any item. Even space requirements are subject to change for some materials and products.

Both shipping and receiving operations are subject to the same general problems and influences. For long-range planning they even share the same major variable: the marketing forecast. However, when a warehouse plan or especially a planning system is developed, the data can be organized in advance of a forecast. Although the specific data needed are unique to each, the planning procedure is fundamentally the same for both the shipping and receiving operations.

12.2.1 Materials Data

A list of the materials that are to be stored and how they are to be stored must be constructed. The planner must establish a list of the end products that are to be stored in the finished goods warehouse. For the receiving area, the planner must construct a list of the raw materials that are to be stored.

In many cases, a complete list of raw materials would be impractical, unwieldy, and, therefore, unusable. In this situation, it would be more practical to study only the raw materials that take up the bulk of the storage space. Out of all the raw materials needed, typically only a few consume most of the storage space. Analyzing these few and factoring for the rest should be sufficient for long-range planning.

Next, a means for converting a marketing forecast into raw materials requirements must be constructed. In some environments this can be established easily through the production

Data Collection   195

control system or a bill of materials. The planner should note
the material usage per end product factored for yield losses.

A relationship between the raw material items to be stored
and the end products to be produced must be established. This
relationship should be established from the bottom up rather
than the top down. Having noted the items that take up the
bulk of the warehouse space, a determination must be made as
to the products with which those items are shipped. The quantity to be stored usually depends on the quantity of product
being produced.

Most manufacturers control their operations through
computer-aided production and inventory control. These systems
usually contain a bill of materials for each end item. In many
cases these systems contain a "where used" query capability.
A component level part number can be entered. The computer
will then list all of the end products where the component is
used and will also print out the quantity of component per end
product.

Of course, with finished goods there is more likely to be
a one-to-one relationship between the item to be stored and the
item or product being forecasted. Unlike determining raw materials requirements, the finished goods usually require no conversion. The end product and a finished good are one and the
same.

Another approach is to contrast current inventory levels
with current production rates. A ratio can then be established
which denotes the quantity of material to be stored per unit
produced within a given time frame. Of course, this approach
assumes current safety stock levels, and inventory turns are
representative for long-range planning.

Whether long-range stores or finished goods stores are
being evaluated, the unit of storage is most important. Typically, the unit of storage is a pallet load. For each finished product and raw material being studied, the planner should note
the quantity per pallet as it is stored in the warehouse. This
documentation must include a notation of the unit's (raw material
or finished goods) per carton as well as the cartons per pallet.

12.2.2

The planner must also determine the space required to store
each material item. Developing the space per storage unit can

be handled in much the same way as developing a footprint for production equipment. One must consider the actual space utilized plus the surrounding space that will be required.

Since most warehouse storage is done by the pallet load on a pallet rack, it is possible to present a standard pallet's footprint (see Fig. 12.1). In addition to the pallet itself, room is required for maneuvering and handling. The footprint must include the room needed to position the pallet within the storage space and for transport to and from the storage space. The footprint, as illustrated, is for a typical warehouse operation. The footprint would be quite different for an automated storage and retrieval system where the aisle widths are narrow.

As demonstrated, each pallet on the floor will account for 5000 square inches or 34.7 sq ft. Any additional pallet loads stored vertically within that same floor space will significantly reduce the square feet per pallet (see Table 12.1). Once the stacking capacity has been determined, the square feet required per pallet stored can be calculated. The square footage per item stored can also be delineated by dividing the square feet per pallet by the items per pallet.

Similar calculations can be made when the material is stored in bulk rather than on racks. Bulk storage can significantly reduce the space required. Bulk storage is usually suitable only when the pallet loads can be stacked on each other without the aid of racks. Also, pallet loads that are stored multilevels deep may create a materials flow problem. Material that is buried in the back row may never be retrieved, creating a first-in last-out materials flow. To avoid this problem, an organized system for materials movement must be maintained.

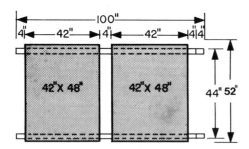

Fig. 12.1 A typical rack which holds 2 pallets would require a footprint of 5000 sq in. per pallet: 52 × 50 in. for the pallet space plus 50 × 48 in. for its share of an 8-ft.-wide aisle.

Data Collection

Table 12.1 Rack Storage: Area Required per Pallet

	Aisle width (ft)							
	5	6	7	8	9	10	11	12
1	28.5	30.6	32.6	34.7	36.8	38.9	41.0	43.1
2	14.2	15.3	16.3	17.4	18.4	19.5	20.5	21.5
3	9.5	10.2	10.9	11.6	12.3	13.0	13.7	14.4
4	7.1	7.6	8.2	8.7	9.2	9.7	10.2	10.8
5	5.7	6.1	6.5	6.9	7.4	7.8	8.2	8.6
6	4.7	5.1	5.4	5.8	6.1	6.5	6.8	7.2
7	4.1	4.4	4.7	5.0	5.3	5.6	5.9	6.2
8	3.6	3.8	4.1	4.3	4.6	4.9	5.1	5.4
9	3.2	3.4	3.6	3.9	4.1	4.3	4.6	4.8
10	2.9	3.1	3.3	3.5	3.7	3.9	4.1	4.3

A footprint for bulk storage two pallets deep is illustrated in Fig. 12.2. Note that the space on the back side of the pallet has been reduced compared to rack storage from 4 to 2 in. This reduction is possible because the pallets can be pushed up against each other. The space on the side of the pallet can also be reduced slightly since there are no uprights, as there are with rack storage. Before making any assumptions, the planner should check a pallet load of each material. In many cases the material may overhang the pallet or the pallet may not be a standard size.

Assuming an 8-ft aisle and a standard pallet, the square footage required per pallet in bulk storage is illustrated in Table 12.2. This figure shows that to store 6 pallets, 3 deep by 2 stacked, 11 sq ft per pallet is utilized. Compared to 2-high rack storage for the same material, a 37% reduction in the required space can be achieved.

The planner should also consider capacity per pallet. There may be instances where bulk storage will enable more cartons per pallet than rack storage or vice versa. Any possibility of improved cubic space should not be overlooked. It could have a significant impact on space requirements planning.

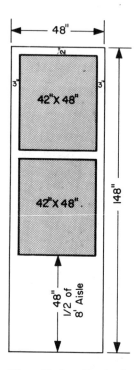

Fig. 12.2 Footprint for bulk storage two pallets deep.

12.2.3 Storage Time

A key ingredient in figuring the space required for warehousing is determining the length of time the material is to be stored. Such factors as safety stock, economic order quantities, and management policy will impact storage time and, therefore, warehousing space, but, by far, the most significant factors are related to inventory turns and inventory fluctuations. The planner must understand their impact on the warehousing operation.

For each material item stored, whether finished goods or raw material, the planner must establish the length of time it is to be stored. In many instances, management will dictate that time. It is not unusual to have safety stock policy and other inventory management goals. The planner should construct a list of each material item stored, how long it is stored, and any

Table 12.2 Bulk Storage: Area Required per Pallet (8-Ft Aisle Assumed)

No. of pallets stacked	No. pallets deep									
	1	2	3	4	5	6	7	8	9	10
1	32.6	24.7	22.0	20.7	19.9	19.3	19.0	18.7	18.4	18.3
2	16.3	12.4	11.0	10.3	9.9	9.7	9.5	9.3	9.2	9.1
3	10.9	8.2	7.3	6.9	6.6	6.4	6.3	6.2	6.1	6.1
4	8.2	6.2	5.5	5.2	5.0	4.8	4.7	4.7	4.6	4.6
5	6.5	4.9	4.4	4.1	4.0	3.9	3.8	3.7	3.7	3.7

safety stock required. The length of time should be expressed in month's worth of inventory, whereas safety stock might be expressed as a fixed quantity. For example, assume a production rate of 1000 units/month. Assume also that 3000 units are usually maintained in finished goods stores. The planner would then note that product's 3-month storage time.

Unfortunately, inventory is not easy to control. Inventory carrying time will vary greatly with product demand and production rates. Any imbalance between what is produced and what is sold will be reflected in inventory. If more of the product is produced than is sold, then finished goods will stay in storage longer. If more can be sold than is produced, inventory will be negligible.

The same situation occurs in the receiving operation. Raw materials inventory reflects the balance between what is purchased and what is produced. In this case, when production slows, inventory may build, and when production increases, inventory will decrease.

Both of these scenarios are created by an imbalance—an imbalance between production and sales or an imbalance between production and purchasing. This fluctuation is not necessarily due to a lack of coordination between these functions. Quite the contrary, in many cases these imbalances are caused by deliberate management strategy. Deliberate fluctuations are particularly prevalent in seasonal industries. However, in any industry imbalances are inevitable which create a corresponding fluctuation in inventory levels.

The planner must be particularly concerned with peak inventories. First, the average number of months of inventory must be established. This length of storage time must then be adjusted to reflect the seasonal peak. Finally, the peak inventory expressed in month's worth of product must be increased to incorporate a possible deviation from the normal peak.

Average Inventory

An average inventory figure can come from a variety of sources, although the two most likely sources are historical analyses and management directive.

Before embarking on an historical analysis of inventory storage time, the planner should consult with management. There may be a stated inventory policy with regard to each finished

Data Collection

good and major raw material. For each raw material, policy may dictate the days, weeks, or months worth which must be in-house. The management of finished goods may be expressed in "inventory turns." Inventory turns refer to the number of times the inventory completely turns over within a year.

Inventory turns are readily converted to the month's worth of storage. For example, four inventory turns mean that there is an average of 3 months of inventory in storage. In an earlier example, it was assumed that a manufacturer produced 1000 units/month, which equals 12000 units/year. If that manufacturer had a goal of four inventory turns, an average inventory of 3000 units would be on-hand. Thus, by dividing the inventory turns into 12 months, the average months of inventory are calculated.

Inventory requirements for each material item are not always dictated by management. The planner may have to extrapolate the inventory data from financial reports, inventory documents, and other historical data. Essentially, the planner must gather two pieces of information. For the past 5 years, the planner should document the annual sales of each product whether in terms of dollars, product units, or pallet loads. If possible, the average annual inventory level of each material should also be gathered for the previous 5 years. From this information, inventory storage levels can then be correlated to production levels. The key to this exercise is to relate storage requirements to the production rate: specifically, the number of months that it would take to produce the finished goods inventory or the number of months it would take to use up the raw materials inventory.

Storage time, in months, can be determined by a number of approaches as illustrated in Table 12.3. While the planner must accept what data are available, it must be converted to compatible terminology. In the example, the planner may know that there were 1000 units produced per month and that $300,000 worth were consistently stored in finished goods inventory. It is then up to the planner to make that information consistent. The units produced can be converted to dollars or the dollars in stores can be converted to units. In either case, the planner must uncover the fact that each unit is worth $100. The planner must know on what criteria the dollar figures are based, i.e., material cost, manufacturing cost, or, perhaps, revenue potential. The planner must be sure to uncover and utilize figures based on consistent definitions.

Table 12.3 Example of Storage/Production/Conversion: Terminology Conversion

Unit of measure	Raw materials (months)	Inventory quantity	Conversion process	Finished goods inventory		
				Quantity	Months	Inventory turns
Product	1	2000 of 1 component	2000 component parts @ 2/product = 1000 products	3000 products	3	4
Pallet	1	100 pallets	100 pallets @ 20 component parts/pallet = 2000 components 2000 components @ 2/product = 1000 products 1000 products @ 4/pallet = 250 pallets	750 pallets	3	4
Dollars	1	$40,000	$40,000 @ $20/component = 2000 components 2000 components @ 2/product = 1000 products 1000 products @ $100/product = $100,000	$300,000	3	4

Assuming production rate = 1000 products/month.

Data Collection

Although this exercise is time consuming, it should be done for each product, or major product group, for each of the last 5 years. This analysis may illustrate a fairly wide fluctuation in the average annual inventory time. These figures should be reviewed with management and a consensus arrived as to what figure is useful for long-range planning. The figure may reflect a goal, but it should also be realistic.

Management may want to set new goals as new methods of materials management are implemented. Perhaps, finished goods will be removed daily from the factory and transported to an off-site distribution center. Finished goods stores in the factory itself might be reduced to only a few hours' worth of production. Changes made in the receiving area could significantly reduce storage time. A "kanban" approach to raw materials management is a good example. This Japanese concept means that materials arrive "just on time."

The data established on average inventory levels are perhaps the major components in warehouse planning. Unfortunately, the warehouse must handle peak levels of inventory, not simply the average levels of inventory. Figures that have been established on average inventory or goals must be factored to reflect seasonal peaks and other deviations from the norm.

Seasonal Adjustment

In most industries, inventory levels are nonconstant throughout the year. The warehouse must be capable of handling the expected inventory peaks. Inventory peaks are usually predictable due to the seasonal nature of most businesses. Seasonal inventory fluctuations usually occur with finished goods only, since it is this inventory that buffers the imbalance between fluctuating demand and consistent production. Raw materials inventory should not fluctuate greatly, assuming the production rate is smoothed throughout the year.

The planner must uncover the seasonal peak for each major finished product to be stored. Peak inventory should be expressed in month's worth of material. Most likely this information will not be readily available. However, for finished goods, it can be extrapolated from the sales history of each product. Preferably, the planner should collect 5 to 10 years sales history by month. It should delineate the number of units sold for each major product line. Utilizing a table of units per pallet, the sales history should be expressed in pallet loads.

For each major product line the number of pallet loads sold per month can be summed as illustrated in Table 12.4. The total number of pallet loads sold by month for the past 5 to 10 years can be used to extrapolate seasonal fluctuations in product demand.

Having established sales totals by month, one can calculate an indexed percentage. An indexed percentage will illustrate the monthly sales fluctuations based on an average of 100%. Each month's percentage of annual sales is calculated by dividing the actual monthly sales history by the annualized average monthly sales. The results of this calculation are illustrated on line 2 of Table 12.4. It shows how each month compares by percentage to an average month of 100%.

Line 3 is simply the difference between the average month of 100% and the actual month's indexed percentage. This monthly variation also tends to show the net effect on inventory. When sales are up, inventory is down and vice versa.

Line 4 illustrates the cumulative effect of this variation. These figures portray the percent of average monthly sales that have been depleted or added to inventory. The result of this cumulative effect is an average level of inventory depletion. In the example, the figures in line 4 result in an average of -0.28. Therefore, due to the cumulative effect, 0.28 months' worth of inventory must be added to bring the average back to zero. This addition of 0.28 is noted as the cumulative factor on line 5.

On line 6, the average month's inventory goal is noted. This goal is the average inventory level (expressed in months) which was established for planning purposes.

The sum of lines 4, 5, and 6 is recorded on line 7. Line 7 represents the inventory level required by month, factored for seasonal variance. In this example, January requires the highest inventory level: 3.43 months' worth. This example illustrates the need to account for 0.43 months' additional inventory to take into account the January sales slump.

Inventory Deviation from the Normal Patterns

In addition to seasonal fluctuations in inventory requirements, storage space may be impacted by other deviations. Deviations from the normal seasonal pattern could be caused by an number of factors, such as the economy, business decisions, vendors, and productivity. Whatever the cause, the warehouse should be capable of housing additional inventory that may be caused by an unpredictable event. Although the event may be unpredictable, its effect on inventory must be planned. A margin of safety can be established by examining historical data. While this approach is not foolproof, it may be helpful as an indication of the range of deviation which could occur.

Projecting what could occur can be extrapolated from what has occurred. In the example, the seasonal nature of the business creates the greatest inventory level in January. Therefore, any deviation that would create an inventory level in excess of January's seasonal peak is of the utmost concern. In this case, the first step is to document the January inventory level's history by product for the last 5 to 10 years. Although sometimes expressed in dollars, it is preferable to obtain the sales inventory levels in terms of units. In either case, the units or dollars must be converted to a relevant common denominator between products.

Since warehousing space is the primary concern and storage is usually done on pallets, this history in units or dollars should be converted to inventory history in pallets. Pallets are a common denominator between product lines. This common denominator will allow the inventory history to be totaled across product lines for each January, as illustrated in Table 12.5. Note also the column that contains the annual sales by pallet load. A ratio can then be established of the January inventory level versus total sales per year.

Once these data have been established, the percentage of deviation from the average January inventory level can be calculated. The highest deviation percentage denotes the worst case that has occurred. In this case, the seasonal peak of 3.43 months would have to be increased by 11.3%. Therefore, 3.82 months' worth of inventory then becomes the calculated highest level of inventory which must be stored.

The results yielded by this exercise must be tempered by sound business judgment. The warehouse should be capable of handling moderate deviation but not a major disaster. Obviously, the company should not construct extra storage space simply to handle a wild deviation in inventory requirements which might or might not occur during the next decade.

There may be some very good reasons why a major deviation occurred and why it should never happen again. The warehouse should be planned to handle some inventory deviation—not necessarily the worst case.

Management should be presented with the facts on past deviations and understand the need to include a contingency

Table 12.4 Product Sales/Inventory Analysis in Pallet Loads (1979 to 1983)

Product	Jan.	Feb.	Mar.	April	May	June	Total
A	1790	2355	2509	2386	2557	2727	
B	3101	4080	4276	4134	4430	4726	
C	6513	8569	8979	8682	9302	9924	
D	1861	2448	2565	2481	2658	2835	
E	2242	2950	3049	2989	3201	3416	
Total pallets	15507	20402	21378	20672	22148	23628	
Average month (%)	85	112	118	113	121	130	
Inventory variation (%)	+15	-12	-18	-13	-21	-30	
Cumulative effect	+.15	+.03	-.15	-.28	-.49	-.79	
Cumulative factor	+.28	+.28	+.28	+.28	+.28	+.28	
Inventory goal (months)	+3.00	+3.00	+3.00	+3.00	+3.00	+3.00	
Inventory level (months)	3.43	3.31	3.13	3.00	2.79	2.49	

Product	July	Aug.	Sept.	Oct.	Nov.	Dec.	Total
A	1694	2203	1511	1762	1735	1965	25194
B	2936	3817	2618	3053	3007	3404	43582
C	6165	8016	5498	6412	6314	7148	91522
D	1761	2290	1571	1832	1804	2042	26148
E	2122	2759	1893	2207	2173	2460	31461
Total pallets	14678	19085	13091	15266	15033	17019	217907/12 = 18159
Average month (%)	81	105	72	84	83	94	
Inventory variation (%)	+19	−5	+28	+16	+17	+6	
Cumulative effect	.60	−.65	−.37	−.21	−.04	+.02	−3.38/12 = −0.28
Cumulative factor	+.28	+.28	+.28	+.28	+.28	+.28	
Inventory goal (months)	+3.00	+3.00	+3.00	+3.00	+3.00	+3.00	
Inventory level (months)	2.68	2.63	2.91	3.07	3.24	3.26	3-month average

Table 12.5 Inventory Deviations from Average Seasonal Patterns (in Pallet Loads)

	A	B	C	D	E	Jan. total	Annual total	Jan. % of annual deviation
Jan. '79	2552	725	492	3254	3255	10278	35200	29.2
Jan. '80	1975	965	396	3714	2691	9741	39123	24.9
Jan. '81	2784	1095	657	4200	5373	14109	43412	32.5
Jan. '82	3150	1132	950	4010	3934	13176	47739	27.6
Jan. '83	3870	1395	1484	4552	5215	16516	52433	31.5
	14331	5312	3979	19730	20468	63820	217907	Average = 29.2

Jan. % of annual deviation
0
−14.7
11.3
−5.5
7.9

Table 12.6 Warehouse Space Profile

	Dallas	Austin	Toledo	Cleveland	Puerto Rico	Miami	Total	%
Total area	30000	80000	82580	114400	25000	10000	341980	
Support areas								
Receiving	1225	5000	6552	5216	3686	2000	23679	7
Shipping	9375	10000	1470	7159	0	0	28004	8
Packing	1225	3000	2100	1096	0	0	7421	2
Returns	0	0	714	1600	0	100	2414	1
OEM Operations	0	0	12960	0	0	0	12960	4
Battery charge	1000	1000	1470	1400	0	0	4870	1
Subtotal	12825	19000	25266	16471	3686	2100	79348	23
Offices	1800	5000	4205	3840	4815	1500	21160	6
Subtotal	14625	24000	29471	20311	8501	3600	100508	29
Storage area	15375	56000	53109	94089	16499	6400	241472	71
Pallet openings	1000	4452	3800	7880	747	465	18344	
Square feet/pallet	15.4	12.6	14	11.9	22.2	13.8	13.2	

factor to handle future deviations. Management should be provided with the opportunity to establish the specific contingency percentage to be used for planning warehouse space.

12.2.4 Support Areas

The warehousing/distribution operation, like the production operation, must allocate space to various support functions. In addition to storage space, there are loading docks, shipping and receiving staging areas, and offices. Space may also be required for some packing operations, returned goods, obsolete item storage, maintenance, and inspection operations.

The current warehousing operation should be analyzed and a record made of space allocation. The percentage of warehouse space allocated to these support areas can then be noted, as illustrated in Table 12.6.

12.3 WAREHOUSING REQUIREMENTS PLANNING

Three factors determine the space required for storage. As outlined earlier, these factors are the quantity of material to be stored, the length of time the material is to be stored, and the space required for storage. Only the first item will vary with a marketing forecast. The others are independent of the forecast and are a part of the base data.

The calculation of storage space for finished goods and raw materials is a straightforward procedure. Although the procedure is fundamentally the same for finished goods and raw materials, the two storage areas should be planned independently for two reasons: (1) raw materials planning requires an extra conversion step, and (2) usually the finished goods and raw materials storage functions are independent both operationally and organizationally.

Warehouse planning is similar to manufacturing planning. The calculations are simple, but the number of calculations is large. Therefore, a properly organized effort will help to ensure accuracy and the efficiency of the planning process. Good organization can be achieved by utilizing a standard spread sheet to tabulate the data and do the calculations.

Warehousing Requirements Planning

12.3.1 Raw Materials Space

The format for developing raw materials storage requirements is illustrated in Table 12.7. Note that much of the planning matrix can be filled in in advance of the marketing forecast (e.g., the parts and their quantity per product, the quantity of parts per pallet, the peak number of weeks of inventory, and the space required per pallet).

Having entered preliminary planning data, one should make a number of copies of this matrix. Copies of the matrix can then be labeled for each time period for which a warehouse plan is to be developed; in this case, a year 2 third quarter plan.

After obtaining a production forecast, the planner then simply enters the weekly rate per product which will drive the rest of the plan. The part number quantity is simply determined by multiplying the parts per product by the number of products. The number of parts divided by the parts per pallet determines the number of pallets per week. The number of pallets per week times the peak number of weeks of inventory will determine the peak number of pallets to be stored.

The number of pallet loads for each component must be multiplied by the space required for each pallet. While some pallets may stack two or three high, other pallets may stack five or six high. Therefore, the floor space required per pallet may be quite different depending upon the item being stored.

The floor space needed to store each product's parts requirements can then be summed. This sum is a projection of the peak floor space needed by those part numbers to support the production plan. However, it is highly unlikely that the part numbers studied reflect a complete list of raw materials inventory. For instance, assume that 10% of the part numbers accounts for 60% of the storage space. Assume also that this plan is based on only that 10% of the parts which accounts for 60% of the storage space. The projected square footage must then be factored in order to account for the 90% of the parts which makes up an additional 40% of the storage space. Dividing the projected space by 0.6 will increase the square footage required to account for the additional space requirements.

The only other factor to be considered is a utilization percentage. Although the peak number of weeks of inventory should account for the worst case, an additional factor for utilization should be considered. A utilization percentage can be

Table 12.7 Format for Calculating Raw Materials Storage Requirements (Year 2 Quarter 3)

Product	Forecast	Part #	Parts/product	Parts forecast	Quantity/pallet	No. pallets/week	No. weeks inventory	No. pallets	Square feet/pallet	Square feet required
A	1200	X1	2	2400	30	80	3.5	280	7	1960
		X2	1	1200	12	100	3.5	350	7	2450
		X3	3	3600	60	60	3.5	210	7	1470
										0
B	240	X2	1	240	12	20	3.5	70	7	490
		X3	5	1200	60	20	3.5	70	7	490

60% of required square feet = 6860. Total storage space = 11433. Total warehouse (2/30% for support) = 16333.

Warehousing Requirements Planning

used to ensure some additional flexibility plus some room for growth. In addition, a utilization percentage must be entered to take into account the warehousing space for other than rack or bulk storage. For instance, 70% of the warehouse may be made up of storage area, while 30% of a warehouse is allocated to loading docks, a supervisor's office, packaging areas, returned goods, and various staging areas. The square footage required for storage must be divided by the utilization factor in order to properly account for these areas.

12.3.2 Finished Goods Space

Determining the storage requirements for finished goods follows almost exactly the same procedure. A finished goods planning matrix is illustrated in Table 12.8. The only real difference is the elimination of the conversion process from product to part number. Finished goods space planning is usually somewhat easier than raw materials space planning. Not only is a conversion step eliminated, but more importantly there are usually much fewer products than there are part numbers. Therefore, with finished goods it may be possible to do a detailed study on the products, which account for 90 to 95% of the finished goods stores area.

12.3.3 Rough-Cut Projections of Head Count and Space

The total space required can be used to extrapolate labor requirements. When projecting the head count to support the warehousing operation, there is usually a linear correlation with the space required. For example, if the current operation requires 50 people for a 200,000-sq ft warehouse, then a 300,000-sq ft warehouse will require roughly 75 people. The problem with this approach is that it assumes that operating conditions will remain unchanged. It also assumes that the average weeks of inventory and the space requirements per pallet remain relatively constant. The problem comes in when a warehouse plan is based on a significant improvement in either of these items. Reducing the space per pallet or the storage time may reduce the size of the proposed warehouse but it will not reduce the number of people. The same quantity of material has to be put on the

Table 12.8 Example of Finished Goods Planning Matrix

Product	Quantity/ week forecast	Quantity/ pallet	Weekly no. pallets	No. of weeks inventory	Total no. pallets	Square feet/ pallet	Total square feet
A	1200	6	200	5	1000	7	7000
B	240	4	60	8	480	7	3360
						Subtotal	10360
						Plus 10% misc.	11396

Warehouse total (@ 70% utilization) = 16280.

Warehousing Requirements Planning

shelf and taken off the shelf, although it may not stay there as long or take up as much space.

A ratio of head count to revenue dollars is perhaps more appropriate for warehouse manpower planning. For instance, a facility that processes $100 million worth of product may have a warehouse staff of 50 people. At $150 million the warehouse staff may have to be increased to 75 people.

Warehousing space requirements can be generated in much the same way. In a pinch or when a crude estimate will suffice, warehouse space can be based on a ratio to revenue dollars. Of course, it is helpful when using this approach to incorporate trends. That is, if the ratio of space to revenue dollars has been decreasing by 10%/year over the last 5 years, the trend must be considered in any calculation of future space. The same is true when a rough-cut head count projection is generated.

13
Manufacturing Tactical Planning and Implementation

13.1 INTRODUCTION

Manufacturing tactical planning is the process of determining the optimal means for meeting the projected requirements. After a requirements plan has been completed, it must be evaluated, revised, presented, and, hopefully, implemented. These four steps are the essence of the tactical planning phase. The objective of this phase is to accurately define and pursue the most effective approach for meeting the projected resource requirements.

13.2 EVALUATING THE REQUIREMENTS PLAN

After a first draft manufacturing plan has been completed, the first priority is to test it for accuracy. Due to the large volume of data and the number of calculations involved, it is highly likely that some errors will occur. The planner must make every effort to rectify the major errors before issuing the results of the manufacturing plan.

There is one technique that is particularly effective for uncovering the glaring errors. Any planner who does not take the time to utilize this technique is asking for trouble. The technique is to simply compare the projected requirements to the present conditions. This simple comparision is a useful technique for uncovering a major problem with the data or a mistake in the calculations.

Future head count, equipment, and floor space requirements should not be wildly out of context with current requirements. For example, in a 5-year plan, assume that most requirements will double. However, in a comparison line item by line item, the planner may notice that a particular piece of equipment has quadrupled. On that line item, the planner should check for a mistake in the calculations or within the data. Simple things can cause significant problems, such as a misplaced decimal point, dividing rather than multiplying by a productivity factor, or, perhaps, a simple error in addition.

Glaring errors are usually easily noticed in the form of an absurd projection of requirements. More often than not, the root of the problem is a significant but simple little error. So, not only is it embarrassing to have a glaring error caught by somebody else, but to have to explain the cause of the error. It is in the best interest of the planner to carefully review and evaluate the projection for each asset requirement and, at least, instinctively test them for reasonableness.

Comparing the projected requirements to the present conditions can place into perspective the magnitude of the new plan. Potential problems with the projected requirements can be recognized and highlighted. For instance, a first pass manufacturing plan may forecast a need for acquisitions that are not really required. Specialized/expensive production equipment or controlled environments may not be acquired until absolutely necessary. Typically, capital-intensive production lines, major equipment, clean rooms, etc., will be run on a two-shift basis before they are duplicated. The initial manufacturing plan might have been based on one shift. In this case, some specific operations or production areas may have to be adjusted to reflect a two-shift operation. The manufacturing plan should reflect the obvious realities of how the company is run.

In some cases, the current assets on hand may be greater than what is projected. In many cases this discrepancy is not because the plan has understated the asset requirements; rather the equipment, labor, or floor space may currently be highly underutilized. However, the plan's projections should be thoroughly investigated and reaffirmed.

Any statement that production management has "too much" is likely to meet with hostility. Shipment goals are more easily met with excess production capacity. Although inefficient, it is in the best interest of the production manager to have the maximum resources possible before appearing inefficient. Any plan that does not justify the resources available is likely to meet with stiff opposition. In essence the plan may suggest that future increases in production can be handled with the existing resources. Such a projection should be well documented in advance of any presentation.

Also recognize that labor and equipment efficiency are not always priorities. In an industry such as electronics or gold processing, it is far more important to keep the material moving and avoid work-in-process inventory. The cost of equipment and labor is negligible compared to the cost of material. Therefore, it may be a deliberate strategy to have "extra" equipment and labor. Such a strategy should be clarified and factored into the plan itself.

13.3 REVISING THE REQUIREMENTS PLAN

Revising the requirements plan, through altering previously "given" operating parameters, is the act of tactical planning. The results from a first pass manufacturing plan may demonstrate the need for a tactical alternative.

The planner should never become attached to any one manufacturing requirements plan. A requirements plan should be viewed as a tool—something to work with. Something must be down on paper in order to facilitate communication between the parties involved. A first pass projection of requirements will enable educated discussion on the allocation and acquisition of those requirements.

13.3.1 Tactical Alternatives

Manufacturing tactical issues involves the allocation of time, location, and method of production. For example, the manipulation of time is done by adding shifts or overtime. The location, or where the manufacturing is done, can be adjusted through the use of vendors, off-shore facilities, or, perhaps, the addition of facilities. Manufacturing methodology, although mostly a strategic issue, may even become a tactical issue that will alter asset requirements. Method issues would include production line design alternatives, such as traditional assembly lines, group technology-work cell production techniques, and horizontal versus vertical integration.

When an alternative approach for meeting the projected requirements is defined, the requirements plan may or may not have to be regenerated. The impact of some alternatives may be readily apparent, while other alternatives require more intensive study. The planner should recognize when a tactical alternative is difficult or easy to reassess.

13.3.2 Time-Related Alternatives

Manufacturing equipment and space is highly dependent upon the available production time. To some extent labor requirements are also influenced by the time available. Available production time can be increased through overtime or by adding a second or third shift. The impact of shifts has already been discussed in the requirements planning section, however, during the strategic planning phase some key points should be kept in mind.

Adding a second or third shift will significantly reduce equipment and space requirements, although it should be kept in mind that additional shifts, particularly the third shift, are not necessarily as productive as the first shift. Also remember that head count requirements are not affected by shifts. It is possible to utilize equipment and space for 16 or 24 hours/day but not people. Running on a two-shift basis may double the work force and double the productive capacity, with no impact on space or equipment.

However, a fractional increase in productive capacity without additional manpower can be met through the use of overtime or additional working hours. An 8-hour rather than a $7\frac{1}{2}$-hour workday is an increase of 6.66%. This increase in productive time applies to equipment and floor space as well as to labor. Also, it should be recognized that an increase in productive time does not necessarily correlate to an equal increase in production. For instance, although a manufacturer pays more for

Revising the Requirements Plan

overtime, actual output may be less than normal due to fatigue. Similarly, an hour of production on the third shift may not be equal to an hour of production on the first shift.

Manipulation of the production time available is a tactical decision. Time is, perhaps, the easiest of the tactical alternatives to evaluate. There is no need to regenerate the entire manufacturing plan. The number of shifts and overtime will affect only the result of the manufacturing plan—labor, equipment, and space requirements. The available productive time does not affect the complex interrelationship between the product and the process.

13.3.3 Location Alternatives

Location alternatives refer to the physical placement of the production operation. Products, subassemblies, or components manufacturing can take place at a number of alternative locations. Items that are currently produced elsewhere can be brought onsite. Conversely, items that are currently produced in-house can be offloaded.

The impact of alternative production locations on manufacturing requirements is not always readily apparent. For example, in many companies only a portion of the products, subassemblies, or components is offloaded to other facilities or vendors, rather than an entire production operation or product line. Outside vendors or off-site facilities may produce some of the items that are also being produced in-house.

Any change in where an item is to be produced will alter the hypothetical production schedule on which the assets plan is based. Therefore, each iteration of an alternative location plan will probably require some recalculation of the manufacturing plan.

In many cases a requirements plan is used to determine which items should be offloaded. For example, a requirements plan may project the need to acquire more production capability than is feasible. The alternative is to have some items produced elsewhere. The requirements plan can help to determine the items that are likely candidates for production elsewhere.

First, a requirements plan will point out the potential bottlenecks which must be avoided. Next, an evaluation can be made of the specific items which, if offloaded, would eliminate the need for more equipment, labor, or space. The labor/equipment hours needed for each operation for each item to be produced are noted on the projection matrix. By manipulating the figures noted in the matrix, location alternatives can be delineated and the impact verified.

Some location alternatives are more easily evaluated. Specifically, the relocation of a process rather than a product/subassembly/component will have a more obvious impact on manufacturing requirements. A requirements plan will usually present the requirements by process, not by product. Therefore, if an entire process is to be offloaded, it is not necessary to manipulate the intricacies of the requirements plan but simply the results of the plan.

13.3.4 Alternative Methods

Usually changes in manufacturing method are not a tactical planning issue. Altering the method by which a product is manufactured goes beyond simply planning future capacity and asset requirements. Methods improvements are more of a strategic design issue.

Methods changes will inevitably be reflected in the industrial engineering standard data. Since these process standards are the foundation of a manufacturing plan, any change in the standard data would require a complete recalculation of the plan. However, with a computer-aided planning system this regeneration effort could be relatively insignificant.

The planning technique, as described in this text, could be used to evaluate the capacity implications of various methods alternatives. However, capacity is only one consideration when analyzing alternative manufacturing methods. Other considerations, not addressed by this planning technique, would include strategic implications, quality, inventory, safety, security, human factors, etc.

Methods changes involve the structure of the manufacturing operation. Changing that structure involves detailed analysis and design. While this planning tool can provide some useful information for the design process, there are other tools better suited to this task. Specifically, there are computer-aided manufacturing simulation software packages available. These systems are designed specifically as an analytical tool, not a planning tool.

13.4 PRESENTATION TO MANAGEMENT

A requirements plan is useless unless it can be communicated and acted upon by others. It is likely that a presentation of manufacturing requirements will have to be done in two formats: (1) an oral presentation and (2) a written report. These presentations are as important as the information itself. A good report/presentation does more than convey information. It should anticipate questions and provide answers. It should recognize problems and provide solutions. Also, any presentation of future manufacturing requirements should instill confidence in the accuracy of those requirements.

Manufacturing requirements are particularly difficult to present either verbally or in writing. Not only must the requirements themselves be adequately displayed, but so must a review of the methodologies, assumptions, and the base data that went into the plan. While the amount of information is overwhelming, its presentation should be well organized and concise.

Usually management is primarily interested in the "bottom line." In particular, they are interested in those items that will necessitate action, especially when the action is needed immediately. Due to the abundance of information involved, it is the responsibility of the planner to highlight the "bottom line." In most instances it is better to present the "bottom line" at the beginning of the presentation/report. Otherwise there is a tendency for people to become impatient and inattentive in anticipation of getting to the bottom line. When presenting a manufacturing plan, whether oral or written, it is the results that are most likely to capture the attention of the audience. Once attention has been captured, it is then possible to effectively present the base data, methodology, and assumptions that were used. By highlighting the bottom line first, there is usually much more interest in knowing how that bottom line was reached. Interest is especially high when the bottom line is shocking, surprising, or controversial.

It is not unusual for a manufacturing plan to be controversial. Because of its comprehensive nature and the number of product- and process-related issues, it is likely that some of the planning parameters will be questioned. It is better to deal with these controversial issues up front. The plan's credibility is enhanced by acknowledging these controversial issues and documenting the related assumptions that were used. In the worst case, the assumptions will be questioned and altered. The manufacturing plan may then have to be regenerated. However, the fact that the controversial operating parameters are not hidden will help demonstrate the integrity of the plan. Management will actually be reassured that the plan is valid since all the facts are "on the table."

In a presentation of future manufacturing requirements, it may be necessary to present alternative strategies and scenarios for achieving those requirements. In many cases, there is not one bottom line but rather a choice among alternatives, e.g., number of shifts or vendor offload. Presenting alternatives is particularly important when the cost/benefit of those alternatives must include evaluation by the financial, business, and executive management. The alternatives are, in effect, the bottom line and should be presented first. The planner can then present the advantages and disadvantages of each from a strictly manufacturing perspective. The presentation should keep within the planner's area of expertise, which is manufacturing. Most likely other considerations such as taxes must be evaluated by others.

Even when one alternative is obvious, it is sometimes beneficial to show other alternatives. A display of these secondary strategies or scenarios could serve two purposes. Only through a comparison of alternatives will the obvious choice become obvious. Second, the thoroughness of the planning study will be demonstrated and, therefore, its credibility enhanced.

The planner should also be prepared to answer "typical" questions. Whether written or oral, the presentation should anticipate some obvious questions. In the presentation of a manufacturing plan, the broad-based assumptions and factors are usually questioned. The planner should be prepared to address two concerns. The first concern is the source of the base data, such as the IE standards, the marketing forecast, and the bill of materials. The second concern relates to the specific factors that were used in the plan, such as the productivity factors that were used for each year, and, perhaps, the offload and second-source assumptions. The planner should also be prepared to document the reasoning and calculations that went into the assumptions and factors.

The planner may wish to document the methodology used to construct the plan. However, a detailed explanation is not usually necessary. In an oral presentation, there is usually not enough time and in a written presentation, it is unlikely that anyone will read it. The key to presenting the methodology that was used is to keep it short and succinct. Usually the planning process can be illustrated on a one-page flow chart. A flow chart can clearly illustrate the procedures that were used without boring the audience. The few individuals who require additional detail can contact the planner directly. In most cases, elaborate presentation of the methodology is not necessary.

The primary emphasis of any presentation should be addressing the concerns of the audience. The audience for a manufacturing plan is simply interested in an accurate projection of future manpower, equipment, or space, but not necessarily all three. One audience may be interested in manpower, while another is more interested in space. Also, some audiences may be more interested in requirements by product line, while other audiences are interested in requirements by facility. In any case, the planner should make every effort to accommodate the primary interests of his or her audience.

Generally, the same rules apply with both written and oral reports. However, each format does contain its own benefits and some idiosyncrasies. In a written report, it is possible to thoroughly document the manufacturing plan. The key to an effective written report is good organization. The planner must recognize that no one is going to read the entire document. Therefore, each reader should be able to find easily those sections that are relevant to him or her.

At the beginning of the written document, an executive summary can be used to succinctly present the plan's bottom line. Most of the audience will read the executive summary but treat the rest of the document as a reference manual. Like any good reference, the planning document should be segmented into clearly labeled sections. A table of contents or index after the executive summary may be particularly beneficial. Perhaps even more helpful are those dividers with tabs that are frequently used in a three-ring binder. In any case, it is important to keep in mind that the written presentation of a manufacturing plan will be used as a reference, therefore, it should be easy to use as a reference.

How the document is segmented is up to the discretion of the planner. However, an effort should be made to best meet the needs of the readers. While the most typical approach is to segment the report by production line, the backup information could be grouped by facility, by year (such as in a 5-year plan), by management organization, etc.

In the written report, every effort should be made to document the manufacturing plan. All of the assumptions, data sources, factors, and to some extent the methodology should be available for anyone to inspect. It then becomes very difficult for anyone to be critical of the bottom line without specifically stating where the planning process is in error. Discussions can then center on specific key issues rather than on general mistrust of the plan.

While an oral presentation does not contain the detail of a written plan, it will usually contain much more clout. In an oral presentation of a manufacturing plan, the response is immediate and, in many cases, the decision makers are present. The planner has an opportunity to request decisions in addition to conveying information. While a written report is purely one-way communication, an oral report can foster two-way communication.

Many times an oral presentation is interactive. At the end of a presentation and, sometimes, during there are always questions. Sometimes, more importantly, there is immediate interaction between the operating managers of the company who have just heard the presentation. Not only is there an immediate response to the plan itself, but also an immediate interchange of ideas and perceptions between the managers present.

The significance of an oral presentation should not be underestimated. An oral presentation should be well constructed and organized, but relatively short and to the point. Time must be left available for discussion.

The planner must appear confident when presenting a manufacturing plan. The presentation should be well organized but not overly polished, nor should it be off-the-cuff. The planner should be sufficiently familiar with the plan to be able to give an oral presentation from a one-page outline, although when presenting trends, forecasts, and the requirements projections, it is helpful to utilize a slide on an overhead projector. It is difficult to present orally a lot of numbers. The first presentation, whether oral or written, is not usually the last. Typically, a manufacturing plan has so many factors going into it that it is inevitable that some things will change. The planner should expect some criticism of and adjustments to the requirements plan. To effectively deal with such changes, the planner should not merely accept revisions, but expect them. There are any number of valid reasons why the planning parameters may change. Management may be privy to some new or sensitive information. Management may also direct changes in the planning parameters

Implementation Planning

simply based on gut feel or experience. These management directives may result in both minor and major revisions of the basic assumptions used in developing the manufacturing plan. Handling these revisions properly will further enhance the confidence of management in the integrity of the planning system.

One of the reasons for having the planning process well organized and documented is to be able to handle changes. These changes present an opportunity to demonstrate the competence of the planner and the speed and accuracy of the planning process.

13.5 IMPLEMENTATION PLANNING

A manufacturing plan is not necessarily complete even after it has been approved. Once the plan is approved, it is highly likely that the planner will be asked to follow through with some sort of implementation plan. Depending on the nature of the plan itself, implementation planning could mean any number of tasks.

The process of acquiring manufacturing assets usually requires budgeting, scheduling, and plant layout. The nature of the plan and the role of the planner within the organization will determine which aspect of implementation planning is the responsibility of the planner.

13.5.1 Budgeting

After settling on what assets are needed, management will most likely want to know how much they will cost. Having a detailed projection of manufacturing requirements should simplify the budgeting process. A good requirements plan will detail the additional labor, space, and equipment requirements. The planner's job will be to determine the cost of acquiring and utilizing the additional assets.

There is one exception. It is highly unlikely that the planner will be required to detail the cost of acquiring additional labor. Most likely that task will fall under the jurisdiction of personnel/employee relations. However, the planner may be asked to estimate the impact of the additional labor on payroll. Usually it will be sufficient to simply multiply the additional head count required by an average annual wage. The result must then be factored to take into account employee benefits including the employer's share of employee-related taxes. It is also important to factor in the projected wage increases. From these calculations a rough estimate of the impact on payroll can be obtained.

A detailed projection of the payroll impact will require further study of the manufacturing labor requirements. The head count requirements will have to be broken down by labor grade. Utilizing the average salary for each labor grade times the number of employees in that labor grade will determine more accurately payroll costs.

It is much more likely that the planner will be asked to determine the cost of acquiring equipment and manufacturing floor space. Fortunately, a detailed requirements plan contains most all of the necessary information. The requirement for additional space can be easily segmented into the three major types: office, manufacturing, and warehouse space. Similarly, equipment requirements are available by specific type and quantity. There are two costs associated with acquiring equipment: the cost of the equipment itself, and the cost of installing the equipment. For each new piece of equipment to be acquired, both of these costs must be determined. In a planning system that is primarily cost oriented, the equipment purchase and installation costs can be incorporated into the planning system itself.

The cost of acquiring additional floor space is sometimes relatively easy to estimate. As for equipment, there are two costs associated with the acquisition of space. There is the cost of obtaining the space and the cost of making it usable. Obtaining the space can be done by leasing, or purchasing, or building. The type of space to be acquired also makes a difference. For any locality being considered, there is usually a standard cost per square foot which can be used for estimating acquisition and set-up cost. Again, both acquisition and set-up costs can be differentiated according to the type of space: manufacturing, office, or warehouse.

A preliminary estimate of the costs involved will usually be sufficient to determine the plan's fate. Assuming that the costs are acceptable, true implementation planning can begin. An implementation plan would include a more detailed budget and a purchase to installation schedule.

A more detailed budget will require discussions and quotes from contractors and vendors. It may even be necessary to go

Implementation Planning

to this level of detail in order to adequately document a capital appropriations request. Detailed budgets and appropriations requests may become the designated responsibility of the manufacturing planner. However, the detailed budgeting/appropriations process goes beyond the scope of this text and therefore will not be covered.

13.5.2 Scheduling

Implementing a manufacturing plan is like any major project. It must be organized and scheduled. The schedule must take into account both lead times and dependencies. In other words, the schedule must take into account the sequence of events, as well as the length of time for each event.

There are a couple of common aids for organizing a complex project schedule: both are charts that illustrate graphically both the timing and the sequence of events. One is known as a program evaluation and review technique (PERT) chart and the other as a Gantt chart.

The Gantt chart or bar chart depicts time on the horizontal axis and jobs or events on the vertical axis. For each event, a horizontal bar is drawn to illustrate the start through completion date (see Fig. 13.1). Not only is the Gantt chart an effective tool for developing and presenting a schedule, but it is also useful for monitoring progress against the schedule. In some

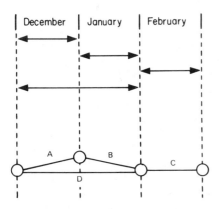

Fig. 13.1 Gantt/PERT chart. (A) Clear area; (B) install utilities; (C) install machines; (D) purchase machines.

cases, the bars on the chart are outlined and are filled in only as the activity or event is completed. Of course, sometimes a bar must be extended to take into account a slip in the schedule. Done with a different color or a separate marking, slips in the schedule become readily apparent. In any case, a Gantt chart is a relatively simple but effective way of formulating, and communicating, any complicated project schedule.

The implementation of a manufacturing plan is typically very complicated. There are many situations where an activity can take place only at the completion of a prior activity. Although a Gantt chart can take into account these interdependencies and relationships, those relationships are not explicitly displayed. However, the PERT chart or network chart is the tool for illustrating graphically these interrelationships.

Below the Gantt chart in Fig. 13.1 is the PERT chart for the same schedule of activities. While the Gantt chart is easier to read, the PERT chart provides more information. A PERT chart clearly depicts what activities must occur prior to other activities.

Another form of network chart is the critical path method (CPM). Actually, CPM is more likely to be used when an implementation schedule is being developed for a manufacturing plan. CPM is more useful where time and cost estimates are available with a relatively high degree of certainty, such as in construction projects. By contrast, PERT is designed to be utilized with research and development projects where there is a lot of uncertainty attached to scheduling time estimates.

In addition to mapping out a network for a chain of events, the CPM approach specifically notes the weakest link. In this case, the part of activities in which the slip of any one activity will result in the slip of the overall plan. Conversely, if the time to complete the entire plan must be shortened, then one or more of the activities on the critical path must also be shortened. In the example, the critical path has been darkened. It is important when utilizing CPM not to overemphasize adherence to the critical path to the exclusion of the other paths. If progress on the other paths is allowed to slip too far, it is possible that a new critical path will emerge and unnecessarily delay the implementation plan.

Any of these scheduling techniques makes it easier to segment and organize the implementation of a manufacturing plan. Having clearly identified the activities, their timing, and their interrelationships, it becomes possible to manage. Each activity can be assigned to an individual or a department depending on

Implementation Planning

the scope of the project. In turn, each individual or department can see the relationship between their portion of the project and the overall manufacturing plan.

13.5.3 Facilities Planning

Timing is not the only consideration in an implementation plan. Not only must the acquisition of manufacturing assets be scheduled, but the location and specific placement must also be planned. There is no sense in knowing when to acquire resources without knowing where to put them. Determining the physical location/placement of manufacturing assets is a necessary component of any implementation plan. Planning the location/placement of resources is generally regarded as facilities planning. This planning could encompass site selection to departmental block plans to quarter-inch scale detail layouts.

Site selection involves determining where to build, purchase, or lease a manufacturing facility. The term site selection is not very specific. A site selection project could mean investigating the relative merits of manufacturing in two different countries. Site selection could also mean determining which plot of land is best within a given city. While the same term may be used to describe both projects, each project is quite different from the other. One is concerned with the general issues of an area; the other involves specific concerns regarding a particular parcel of land. However, in either study, the same issues must be addressed, but at a different level of detail.

Whether evaluating a general local or a specific site, there are six major concerns that are common to both. Site selection must consider the impact of government, business, transportation, labor, utilities, and quality of life.

In evaluating a specific site or plot of land, there are some additional concerns not applicable to a general location study. Specific site considerations would involve topography, size, orientation, drainage, flooding potential, subsoil, excavating, and landscaping considerations.

All site selection studies should address the following factors.

1. Government. The influence of government on site selection can be substantial. Governments that are seeking to attract manufacturers may provide inducements in the form of grants, financing arrangements,

or tax breaks. Another concern is the impact of all applicable taxes. Similarly, there may be a number of regulations that can impact a manufacturer's ability to do business. In some situations, the stability of the government itself is a legitimate concern. In all cases, the availability and quality of government services such as police, fire, and hospitals should be carefully evaluated.
2. Business. The availability of other businesses may be essential to an effective manufacturing operation. In most cases, raw materials will come from other businesses. The availability and cost of raw materials are typically primary considerations. Other businesses may also be necessary to supply industrial and business services. Industrial services would include maintenance and repair for the facility as well as for the equipment. Business services might range from consultants and lawyers to testing laboratories or printers.
3. Transportation. Any manufacturing facility depends on the ability to ship and to receive. The infrastructure that surrounds a manufacturing facility may determine how well these functions are performed. Roads, airports, railways, and waterways could be major considerations in the selection of any site.
4. Labor. A primary consideration is labor, its availability and cost. Other considerations that could be critical are the skills or education level of the work force and their attitude toward work. The availability of management, technical, and professional labor also should not be overlooked.
5. Utilities. Again availability and cost must be considered. Energy, whether in the form of oil, gas, or electricity, is a major consideration. Water is also essential. Communication services may also be important.
6. Quality of life. This category contains a number of factors that should be considered in the selection of a site. Such factors include the weather/climate, the culture, cultural activities, recreational facilities, housing, outdoor attractions, shopping, etc.

Site selection involves an endless number of considerations. Each company has its own objectives and requirements for a site.

Implementation Planning

Therefore, it is impossible to develop a site selection checklist that is universally applicable. Each company should have its own checklist that emphasizes the company's concerns with the six major factors.

Whether developing a long-range plan for a proposed site or an existing facility, plant layout plans will eventually be necessary. Long-range layout planning is usually done in two stages. The first stage is to develop an overall block plan for the plant being studied. The second stage is to develop a 1/8- or 1/4-in. scale layout showing equipment locations, storage areas, aisles, etc.

Prior to detail planning, a block plan should be done to show the relative size and location of various production line departments or work centers within the manufacturing plant. In the design of a new plant, the blocks may determine the size and shape of the plant. However, in planning the layout of an existing facility, the facility may determine the shape of the blocks.

In an existing facility, the permanent fixtures should also be outlined in the block plan. Fixtures such as elevators, stairwells, columns, windows, doors, and fixed aisles should be clearly displayed. In some cases, there are physical constraints that will dictate the location of a particular manufacturing department within the plant. For example, some departments may require access to existing drainage trenches, while other departments may require access to existing clean rooms. When designing a block plan, the planner must be sure to note first those blocks that are fixed by the facility itself.

Another fundamental consideration in block planning is material movement. For instance, material movement to and from a warehouse can usually be measured in pallet loads per day. Those manufacturing departments with the highest number of pallets moved per day are the most likely candidates for location nearest to the warehouse. Conversely departments with minimal flow of materials do not have to be near the warehouse.

In some situations, it is equally important to consider a manufacturing department's access to various auxiliary or support departments. For example, some manufacturing departments may be the company's showpiece. They are the essence of the "factory tour." In some companies, it may be essential to place such a department near to the sales offices or executive suite. Similarly, there are a number of other areas that may help dictate the location of a particular department on a block plan.

The location of some departments may be dictated by their proximity to research and development laboratories, washrooms, cafeterias, medical clinics, accounting offices, maintenance/machine shops, engineering offices, etc.

After a block plan is established, it is possible to construct a detailed 1/8- or 1/4-in. scale plant layout. However, detailed layouts are usually only done, or needed, for short- to medium-range manufacturing plans. A detailed plant layout will illustrate the location of each piece of piece of equipment, material handling devices, aisles, and storage locations. Most detailed plans should also allow room for expansion and some space for the unexpected.

During the design phase of a new plant layout, every effort should be made to include productivity planning. The implementation of a new plant layout provides an ideal opportunity to improve the existing method of operation. The design of a new plant layout presents the opportunity to improve the flow of materials, improve how the materials are transported, and improve the methods by which the product is produced. However, no improvements should be incorporated into the manufacturing layout plan unless great care and study have gone into their feasibility. Although any methods change involves risk, the confidence level for success should be fairly high. The odds on success should be extremely high when the new layout will preclude a reversion to the old method. For example, in process- or capital-intensive industries, the installation or relocation of equipment may be extremely expensive. A major layout/methods "improvement" that does not work, and has to be reversed, could be devastating.

In addition to the final plant layouts that may be necessary, there may also be intermediary layouts that are necessary. There may be a need for time-phased implementation layouts. In order to achieve the final layout, a number of steps may be required over a period of time. A separate plant layout may be necessary for each phase of the total implementation plan. In many cases, the new plan's logistics are so complex that they will necessitate an illustration of each step.

14
Computer-Aided Manufacturing Assets Planning

14.1 FEASIBILITY ANALYSIS

Computers can be extremely helpful and sometimes essential for the development of a manufacturing plan. Utilizing a given set of operating parameters, a computer can project manufacturing requirements. For the most part, a computer can generate requirements, but not determine strategy. A computer-generated requirements plan can be used as a feedback mechanism for others to determine strategy. A computer can do the requirements planning phase but not the strategic planning phase of the manufacturing planning process.

The feasibility of utilizing a computer hinges upon the objectives that have been established for manufacturing planning. Also, the complexity of the manufacturing operation will help to determine the need for a computer.

The objectives for manufacturing planning must be documented specifically in terms of what information is desired and when it is desired. A definition of information needs can be expressed in terms of accuracy, level of detail, presentation format, etc. The timing of the information can refer to both the frequency and the response time.

Primarily, the objectives must be defined for what information is desired and when it is desired. The information desired must be expressed in more detail for a computer-aided system than for a one-shot manufacturing plan. The level of detail desired must be defined precisely.

The main purpose of any manufacturing planning system is to provide information in the form of a printout. Therefore, the design of the output report will focus direction for the entire planning system. Not only will each piece of output information be specified, but it will also help clarify the inputs that are needed.

The output report should be designed in total. As a mock-up of a typical report it should include everything just as if it came out of the computer: the headings, the information required by operation, by work center, by department, and by product line, and an example of the summary report.

In addition to the information itself, some determination must be made as to the degree of accuracy desired. For example, will the system use estimated standards for part groupings or the actual standards for each part number? Again, the planner may have to mediate the conflict between timeliness and accuracy.

The timing of the delivery of the information is also an acute consideration. When the system's objectives are documented, the expected or desired frequency of use should be established. It is necessary to recognize the number of times per week, per month, or per year that the system will be utilized to generate a requirements plan. Be careful to include the number of times that the parameters are altered and a new plan is generated. For instance, if the system can handle only one period at a time, then it will take five separate runs to generate a 5-year plan. Also, it is necessary to consider the number of times the system will be run for correcting mistakes, changing parameters, and alternating scenarios and strategies.

In addition to frequency, there may also be an objective in regard to the speed with which the system can deliver the information. In some cases, the interval between the availability of a new marketing forecast, and the delivery of the manufacturing requirements report, must not exceed a specified period of time. Information is valuable only when it can be used.

In addition to the initial response time, more time may be needed for subsequent reiteration of the manufacturing plan.

Feasibility Analysis

The ability to generate a new plan based on alternative vendor and production strategies may be crucial. The main benefit of a computer-aided planning system may be the ability to answer "what if" scenarios. In essence, the objectives are a statement of what is desired from the planning system. However, the viability of a computer-aided system also depends on the environment or operating parameters with which the system must work. Since the system's function is to gather and process data, the data become a necessary parameter with which the system must operate. Again, the objectives are a statement of what is desired from the planning system, while the operating parameters are the data requirements needed to achieve those objectives.

The feasibility of a computer application depends upon the complexity of the data, the volume of the data, and the accessibility/availability of data on computer. The complexity and volume of data may help determine the need for a computer, while the accessibility/availability of computerized data will determine whether or not it is even possible in the short term.

The volume of data that must be manipulated is probably the most significant determinant of the need for a computer. In a large factory with many products of complex structures, the bill of materials data could be huge. Similarly, the engineering standards for each component could represent a huge volume of data. In any case, the volume of data required to meet the planning system's objectives must be understood and documented.

The complexity of the data is also important. Complexity becomes an issue when there are a number of calculations needed to get it into the proper format. Data gathering and manipulation by computer also become complex when the data are already on a computer file but it is difficult to read.

The accessibility/availability of data is a major concern when a planning system is being developed. In most cases, if the major data files (bill of materials, i.e., standards) are either unavailable or inaccessible, then there is no sense in developing a computer-aided planning system around them. This does not mean that computer-aided planning cannot be done; it does mean that the planner should not expect major data files to be created to support a manufacturing planning system. In the absence of existing files, the planner may improvise. However, everyone involved must recognize that the development and maintenance of any major file may sometimes become a huge project in itself. Whenever possible, the planner should avoid the development and maintenance of data files. A planning system will work best when the data files are maintained and used regularly by others.

Having analyzed the outputs desired and the inputs required, the feasibility of computer-aided planning usually becomes readily apparent. In many situations not only is a computer-aided system feasible, but it is an absolute necessity. In these cases there is no way to achieve the desired results without using a computer. On the other hand, there are cases where it becomes readily apparent where the use of a computer would be an overkill. For some companies, manually generated manufacturing plans represent the most reasonable approach.

The feasibility of utilizing a computer is entirely dependent on the relationship between what is desired and what it will take to meet that desire. The number of manufacturing plans per year and the number of calculations required per plan will determine whether or not a computer is needed. The need to utilize a computer becomes apparent when either the plans must be generated very frequently or there are a large number of calculations per plan. The use of a computer may also become feasibile in a situation of moderate complexity with moderate frequency.

14.2 REQUIREMENTS ANALYSIS

Manufacturing planning is usually an excellent application for a computer system. In many cases, detailed manufacturing planning is nearly impossible without the aid of a computer. Having determined that the use of a computer is feasible, one must determine the required computing capability.

Although interwined, the computer requirements can be broken down into hardware requirements and software requirements. Both depend on the amount of information being processed, how often it is being processed, and the turnaround time desired. The same factors that were used to determine feasibility must be investigated further in order to determine the required computing capability.

There is no right computer for manufacturing planning. Like individual manufacturing plans, planning systems vary greatly between industries, companies, etc. Computing power must be based on the specific needs of one's business and proposed planning system. Ideally, the computing system should

Requirements Analysis

be tailored to match the needs of the job—not the other way around.

Primarily, it is important to document the processing and data requirements of one's proposed planning system. The planning system, like any computing system, will consist of inputs and outputs. The inputs and outputs must be described along with the programs needed to make the conversions.

Fig. 14.1 illustrates the inputs and outputs of a typical manufacturing planning system. Information is stored on the computer as files. In this case, there are forecast files, which consist of the product forecast, the spares/offloads forecast, a subassembly forecast, and a forecast of productivity indices.

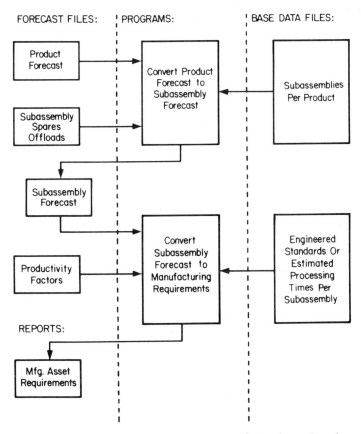

Fig. 14.1 The computer-aided manufacturing planning process.

These forecast files are entirely related to the time frame being studied. Unrelated to the forecast period is the planning system's base data. In this case, there are two major base data files: the bill of materials file and the engineering standards file. These two files are usually very large and the primary factor in determining computer needs. The planning system must also include two major programming modules and an output report. The planning process can be broken down into three categories of information: variable data, programs, and base data.

The main purpose of the flow chart is to segment the planning process into manageable entities. It is possible to evaluate the information needs of each module of the planning process. It is also important to recognize the advantages of this system's design. Its major attributes are that the system is modular and that data are kept separate from programs. The programs merely manipulate files that contain all the information. Therefore, the programs are generic, i.e., they can be used for any manufacturing plan simply by altering the files that are accessed.

From a programming standpoint, the software requirements are relatively minimal. The programming must do two tasks: manipulate files and crunch numbers. The programming that is needed to do these simple tasks should consist only of a few hundred lines of code. However, a couple of thousand lines of code may be needed to make the program "user friendly."

Manufacturing planning system software could be written in any one of the major programming languages. In many situations, there is no choice as to which programming language to use. Frequently, the person available to do the programming knows only one programming language. That person may be the planner who happens to know some programming.

If there is a choice of programming languages, it will probably be among BASIC, FORTRAN, and COBOL. The relative merit of each of these languages is noted below.

> BASIC. When a planning system is established, there are two problems. First, BASIC is not very basic. There are numerous versions of BASIC each with its own idiosyncrasies. Second, many versions of BASIC require commas between the data elements of the data files. This constraint could preclude the use of existing data files.
>
> FORTRAN. Designed primarily for scientific applications and complex algorithms, FORTRAN is quite workable for

Requirements Analysis

manufacturing planning. FORTRAN is universally accepted and standardized. More important, FORTRAN is easily programmed to read any existing file, assuming that there is consistency in the location of data elements within that data file.

COBOL. This business/data processing language is also applicable to manufacturing planning systems. It is capable of handling organized files of data and doing the number crunching. The only drawback to COBOL is that it is very wordy. It will take more lines of code to do a task in COBOL than it will take to do the same task in BASIC or FORTRAN.

Between FORTRAN and COBOL, it is probably best to use the language most familiar to the person doing the programming. Each language is capable of accomplishing the same end result. Of course, software is also related to the hardware on which it is being run. The computer that is available may determine which software package to use. For instance, some computers that are capable of running all three languages may do only one very well.

For manufacturing planning, the major issues involving hardware are processing capacity and storage capacity. Processing capacity will primarily determine the speed with which a manufacturing plan can be generated. The amount of information needed to generate a plan will determine the nature of the storage devices that are required.

A computer's processing capability is determined by the type of central processor. The type of processor is usually discussed in terms of bits and random access memory (RAM). For instance, there has recently been much discussion on the relative merit of 16-bit microcomputers versus 32-bit mini-computers. While both are very capable business machines, the same program will take longer to run on a 16-bit machine. If speed is a priority and the volume of data is large, then the power of the machine expressed in bits could be a significant consideration.

A computer's processing capacity is also influenced by its internal memory or RAM. However, for manufacturing planning, RAM should not be a significant consideration. Old business machines have a minimum of 64K RAM, which is sufficient for most manufacturing planning programs. Of course, this assumes that the programs are written efficiently and minimize the utilization of internal RAM.

Probably, the major consideration for manufacturing planning systems is a computer's storage capacity. This capacity is usually measured in bytes. A byte is usually equivalent to a letter or number. From this simple fact, it is possible to determine the amount of storage space that will be required.

An estimate must be made of the number of characters to be contained in each file. When counting the number of characters, be sure to include spaces. A space takes up as much room as a letter or a number. Also recognize that an estimate is all that is needed. Assume that the bill of materials file is expected to contain 60 characters of information per line and be 1000 lines long. In computer jargon, the size of this file would be 60K bytes. If the IE standards file and the bill of materials file each contain 60K bytes and all of the other files amount to 40K bytes, the total requirement for storage would be 160K bytes. Before purchasing a computer with exactly 160K bytes of external storage, the planner should be sure to include a growth factor. Both the bill of materials and the standards files could grow significantly. Therefore, it might make sense to look for a computer with 300K bytes of external memory.

However, it is more likely that a manufacturing plan will be established on existing equipment. Beyond availability, there are some other good reasons for utilizing existing equipment.

14.3 UTILIZING THE "IN-HOUSE" MAINFRAME COMPUTER

The use of a major "in-house" computer has a number of advantages for manufacturing planning, the primary advantage being that it is available. More time can be spent installing the planning system rather than investigating and justifying a new computer system. Typically, the acquisition of new computer equipment is a project in itself. In most cases, the acquisition of new computer equipment is not necessary since an available company computer is usually more than sufficient for manufacturing planning.

Most in-house business computers are relatively large. Not only is their central processing ability extensive, but so is their external storage capacity. In comparison to other major business software packages, a manufacturing planning system can be relatively small.

Utilizing the "In-House" Mainframe Computer

By far the most significant attribute of the in-house computer is the availability of data. In fact, most of the manufacturing planning data may already exist and be maintained by others. Accessibility to existing data bases is a major asset. It could mean the difference between the long-term success or failure of a manufacturing planning system.

Access to maintained data is the only way a large-scale manufacturing planning system will work. It is impossible for a planner to maintain a bill of materials file and a standard data file, except perhaps in a very small factory. However, by designing the planning system around existing data files, the planning system can become self-sustaining. It will always be ready to run with the most recent data.

The in-house computer can also provide access to the manufacturing planning system by others. For instance, specific production departments may wish to utilize the system for a projection of next week's or next month's capacity constraints, or perhaps the financial department will automatically access the projection of equipment requirements in order to generate a capital plan. Another significant advantage of the in-house computer is that it is supported. System analysts, programmers, and other technicians are available. Although typically overburdened, the management information systems (MIS) department is there to support the computer system. Support usually consists of maintenance and enhancements to both the hardware and the software. This support can be a major asset when trying to get a manufacturing planning system up and running.

However, this "support" can sometimes be a disadvantage. Both hardware and software are under control of someone other than the planner. Their priorities do not necessarily lie with manufacturing planning. It is sometimes very difficult to obtain programming support. Perhaps worse, a perfectly operational manufacturing planning system could be bumped off the computer to make room for other jobs. In either case, the ability to generate a manufacturing plan is hampered by others beyond the planner's control.

In most cases, the availability of maintained data far outweighs any loss of control. If the system is properly implemented, the loss of some control should be of minimal concern. The planning system should still receive the priority deemed necessary by management.

When the planning system resides on the in-house mainframe computer, collaboration with the MIS is unavoidable, although this collaboration could range from simply obtaining an

ID and password to obtaining programming assistance. In either case, knowing what to ask for from the MIS department can facilitate understanding and cooperation.

In the easiest case, the planner may simply be asking for access to the computing system. Even so, the MIS department may need to know how the system is going to be used and how often. The planner may have to specify how much storage capacity is needed for manufacturing planning. For instance, if the major data files are already on the system, then the planning system itself will require very little additional storage.

As with any user, the planner can be expected to put a demand on the computer's time. The MIS group may need some estimate of how much time the planner expects to spend on the computer. They may also need to understand the type of programs being run along with the size of the files being accessed and the complexity of the calculations being performed.

14.4 MICROCOMPUTER APPLICATIONS

The proliferation of microcomputers as a management/engineering tool can provide an almost immediate capability for manufacturing planning. Controlled by an engineer or manager, a microcomputer eliminates the dependency on any outside group, such as MIS. Through the use of readily available "spread sheet" software packages, the planner can easily program his or her planning system.

There is one major drawback to using a stand-alone desktop microcomputer. That drawback is the difficulty of updating and maintaining an accurate data base. However, in the near future improvements in hardware and software technology will enable microcomputers to access and utilize major data bases that are stored on larger systems.

In some industries, this drawback is of minimal consequence. In industries where the product and process remain relatively stable, the supporting data base should also remain stable and require minimal maintenance. Therefore, a large volume of data may have to be entered into the computer only once. Once it has been entered, the data base may remain relevant for months or even years. Conversely, in high-technology, rapid-growth industries, the product and process are also undergoing rapid change and along with them the product/process data base.

Microcomputer Applications

Assuming that the relevant data are manageable, not too large, and relatively static, then a microcomputer may be more than sufficient for manufacturing planning. There are two approaches to programming a planning system on a microcomputer. One approach is to develop it in exactly the same way one would develop it on the in-house mainframe computer, using data files and a standard programming language such as BASIC. The programmed manipulation of files is the same whether done on a mainframe or on a microcomputer. Therefore, this approach will not be discussed further. The alternative approach that has not been discussed is to utilize a spread sheet program.

A spread sheet program allows the easy manipulation of rows and columns of data. When purchased, a spread sheet program is generic in nature. The spread sheet is blank with no preconceived structure other than a blank matrix with the ability to manipulate that matrix as the user sees fit. Spread sheet programs are applicable in almost any business situation where calculations were previously performed on ledger paper with ruled columns and rows.

Like other business analysis, manufacturing planning can be done on a spread sheet. However, like any other business study, revisions are inevitable. Whenever one piece of data or box in the matrix is changed, a large part of the entire matrix may have to be recalculated. Done manually on ledger paper, such a recalculation could take hours; executing the same recalculation on a microcomputer could take minutes. A spread sheet program can provide instant feedback to a number of different operating scenarios and conditions.

One problem in developing a manufacturing plan by hand is that when it is completed someone will ask, "What if?". Done manually, these questions might not be worth the time and effort required to answer them. Done with a microcomputer, "What if?" can be answered with very little effort. The value of this rapid feedback capability should not be underestimated. Rather than a one-shot static report in a three-ring binder, manufacturing planning can become a dynamic interactive management tool.

Essentially, spread sheet programs work by assigning each cell in the matrix its own distinctive code name. For example, if the columns are denoted as A B C . . . and the rows 1 2 3 . . ., then the first cell in the matrix—the upper left hand corner—would be labeled A1.

In addition to a code name for each cell, the cell can contain a piece of data such as a number entered by the planner and its own formula. The formula that is attached to a cell allows the data within that cell to be calculated. In other words, the data contained within a cell may be the result of calculations performed on data gathered from other cells. For instance, if column A contained the number of employees within a factory and column B contained the square footage within a factory, the formula for column C could be column B divided by column A or square foot per employee. In any case, it is up to the user to decide which cells are to contain data and which are to be assigned formulas.

The size of the spread sheet can be limited by either the design of the software or the capacity of the hardware. Before pursuing the spread sheet/microcomputer route, the planner should check and see if these limitations are too restrictive. For example, VISICALC, the father of the spread sheet programs, can handle 63 columns across and 254 rows down— roughly 16,000 cells. However, many microcomputers cannot accommodate that many cells. Therefore, before embarking on the spread sheet/microcomputer approach, the planner must at least conceptually design and determine the size of the proposed planning system. In other words, design the application and then acquire the software and hardware to do the job.

Spread sheet programs can be extremely useful for manufacturing planning in any company. Because of the unlimited versatility of spread sheet programs, there is no one correct way of setting up a manufacturing planning system. Spread sheet programs can accommodate everything from a rough-cut planning approach to a more detailed and sophisticated analysis.

Table 14.1 is a rough-cut macro planning model utilizing a typical spread sheet program. It models the entire manufacturing operation for a Fortune 500 firm. The system projects head count and manufacturing space. The system utilizes a ratio/trend approach, based on revenue dollars per square foot and revenue dollars per employee.

In this case, 1982 is considered the base year. Output data were quantified for each manufacturing/product line as of year end. Departments that do not generate revenue, such as administrative areas and research and development areas, were ratioed to total revenue dollars for the corporation. Also note that output for the same product is affected by where it is produced.

The system works by entering the total revenue dollars for any year being studied, in this case 1985. The revenue is

Microcomputer Applications

then divided among each manufacturing department according to the percentages entered in column 1. A different year, such as 1986, is done on its own spread sheet with revenue percentages allocated according to its own forecast.

Columns 3, 4, and 5 contain the fixed 1982 profile of production ratio data. The figures have been calculated at capacity, not actual. In other words, the figures are based on what output the lines are capable of doing, not on what they actually did. It was necessary to calculate the output at capacity since some newly implemented production lines were producing a small fraction of their designed capability. However, expecting 100% utilization of capacity is unrealistic, therefore column 6 denotes a more reasonable utilization percentage.

Column 7 shows the productivity improvement in output capacity per square foot. An across-the-board improvement of 41% is expected, which means that it takes 59% of the 1982 space to produce the same output. The system divides column 3, the 1982 output per square foot, by column 7 (0.59), to calculate column 9, the projected revenue per square foot for 1985. Note that although "utilization" and "productivity" figures are the same for each manufacturing department, the system is capable of handling separate figures for each department.

Column 10, "Revenue Dollars per Employee," has been calculated much the same as for column 9. The main difference is that the productivity improvement percentage for labor output is invisible to the spread sheet (although it did not have to be). Labor output was calculated, internal to the program, at 10% per year; specifically, 1985's labor should be 72% of what it would have taken in 1982.

Column 11 presents the head count projection for each manufacturing department. Simply stated, "head count" is calculated by:

$$\frac{\text{Projected revenue (column 2)}}{\text{1982 Revenue per employee (column 4)}} \times \text{Labor productivity factor}$$

= Projected head count

Column 12, the "square footage," is calculated from the following formula:

Table 14.1 Rough-Cut Manufacturing Plan

Product	Plant	Revenue (%)	K Revenue ($)	$/sq ft	$/employee	Sq/ft employee	Utilization
A	Dallas	16.33	430949	1226	337143	324	85
A	Puerto Rico	0.37	9764	421	209524	585	85
B	Dallas	10.45	275776	1335	334270	295	85
B	Puerto Rico	2.17	57266	514	219298	502	85
B	Buffalo	1.68	44335	618	185577	354	85
C	Austin	13.45	354946	2998	1870000	734	85
C	Puerto Rico	0.55	14515	438	217143	584	85
D	Toledo	0	0	292	122137	492	85
D	Puerto Rico	0	0	479	244444	600	85
E	Toledo	31.2	823368	1253	378919	356	85
E	Puerto Rico	1.8	47502	610	300000	578	85
F	Dallas	2	52780	647	214667	390	85
G	Toledo	16	422240	997	221651	262	85
H	Austin	4	105560	1612	376106	275	85

I	North Carolina	6.96	183674	639	400000	736	85
J	North Carolina	2.09	55155	313	195122	733	85
K	Toledo	0.52	13723	625	181818	342	85
L	Buffalo	2.39	63072	1599	158960	117	85
L	Lincoln	0.04	1056	351	27778	93	85
M	Lincoln	0.54	14251	351	27803	93	85
N	Lincoln	0.1	2639	341	27273	94	85
N	Toledo	1.96	51724	1600	158451	116	85
O	Lincoln	0.52	13723	3324	272727	97	85
P	Park City	100	2639000	108286	26136000	284	85
Q	Park City	0.55	14515	56	28899	603	85
R	Austin	0.43	11348	103	24752	282	85
Distribution	Dallas	93.5	2467465	7241	3258485	500	90
Distribution	Puerto Rico	6.5	171535	3458	1114925	358	90
Manufacturing administration	Manufacturing administration	100	2639000	7416	1218220	183	90
Distribution center	Dallas 2	100	2639000	30066	13372093	523	85

Table 14.1 (Continued)

Product	Plant	Productivity	Sq ft/ employee	$/sq ft	$/employee	Head count	Sq/ft sum
A	Dallas	59	265	1766	468254	920	244056
A	Puerto Rico	59	479	607	291005	34	16081
B	Dallas	59	241	1923	464263	594	143383
B	Puerto Rico	59	411	741	304581	188	77274
B	Buffalo	59	290	890	257746	172	49836
C	Austin	59	601	4319	2597222	137	82173
C	Puerto Rico	59	478	630	301587	48	23026
D	Toledo	59	0	421	169635	0	0
D	Puerto Rico	59	0	691	339506	0	0
E	Toledo	59	292	1805	526277	1565	456095
E	Puerto Rico	59	474	879	416667	114	54018
F	Dallas	59	320	932	298148	177	56618
G	Toledo	59	214	1436	307849	1372	293966
H	Austin	59	225	2322	522370	202	45463
I	North Carolina	59	603	921	555556	331	199517
J	North Carolina	59	601	451	271003	204	122314

K	Toledo	59	280	252525	54	15240
L	Buffalo	59	96	220777	286	27374
L	Lincoln	59	76	38580	27	2088
M	Lincoln	59	76	38615	369	28191
N	Lincoln	59	77	37879	70	5366
N	Toledo	59	95	220070	235	22435
O	Lincoln	59	79	378788	36	2866
P	Park City	59	233	36300000	73	16916
Q	Park City	59	494	40138	362	178631
R	Austin	59	231	34378	330	76280
Distribution	Dallas	59	410	4525674	545	223389
Distribution	Puerto Rico	59	294	1548507	111	32519
Manufacturing administration	Manufacturing administration	59	150	1691973	1560	133281
Distribution center	Dallas 2	59	429	18572351	142	60926
Subtotal					10256	2789320
Utilization (%)						—
Manufacturing (%)						—
Forecasted						3080103
Current						529500
Additional required						−449397

Table 14.1 (Continued)

Product	Plant	sq/ft sum	Dallas	Toledo	Park City	Cleveland
A	Dallas	244056	244056			
A	Puerto Rico	16081				
B	Dallas	143383	143383			
B	Puerto Rico	77274				
B	Buffalo	49836				
C	Austin	82173				
C	Puerto Rico	23026				
D	Toledo	0				
D	Puerto Rico	0				
E	Toledo	456095		456095		
E	Puerto Rico	54018				
F	Dallas	56618	56618			
G	Toledo	293966		293966		
H	Austin	45463			45463	
I	North Carolina	199517				
J	North Carolina	122314		105557		
K	Toledo	15240				

L	Buffalo	27247			
L	Lincoln	2088			
M	Lincoln	28191			
N	Lincoln	5366			
N	Toledo	22435			
O	Lincoln	2866			
P	Park City	16916		16916	
Q	Park City	178631		158624	
R	Austin	76280			
Distribution	Dallas	223389	223389		
Distribution	Puerto Rico	32519			
Manufacturing administration	Manufacturing administration	233281	223281		
Distribution center	Dallas 2	60926			
Subtotal		2789320	900727	870858	2211003
Utilization (%)		—	100	100	100
Manufacturing (%)		—	100	99.4	72.4
Forecasted		3090103	900727	876115	305252
Current		529500	739000	782100	209000
Additional required		−449397	161727	94015	96252
					167000
					−167000

Table 14.1 (Continued)

Product	Plant	Austin	Dallas 2	Hartford	Boston	Buffalo	Puerto Rico
A	Dallas						16081
A	Puerto Rico						
B	Dallas						77274
B	Puerto Rico	82173				49836	
B	Buffalo						
C	Austin						
C	Puerto Rico						23026
D	Toledo						
D	Puerto Rico						
E	Toledo						
E	Puerto Rico						54018
F	Dallas						
G	Toledo						
H	Austin						
I	North Carolina						
J	North Carolina						
K	Toledo						
L	Buffalo					27347	

		C1	C2	C3	C4	C5
L	Lincoln					
M	Lincoln					
N	Lincoln					
N	Toledo				22435	
O	Lincoln					
P	Park City					
Q	Park City					
R	Austin	76280				
Distribution	Dallas					
Distribution	Puerto Rico					32519
Manufacturing administration						
Distribution center	Dallas 2		60926			
Subtotal		158453	60926		99646	202917
Utilization (%)		100	100		100	100
Manufactuirng (%)		100	100		100	100
Forecasted		158435	60926		99646	202917
Current		123100	56200	203100	300000	320000
Additional required		35353	4726	−203100	−200354	−117083

Table 14.1 (Continued)

Product	Plant	Lincoln	Dallas 3	Singapore	Canada	Canada 2	North Carolina	Austin 2	Ireland
A	Dallas								
A	Puerto Rico								
B	Dallas								
B	Puerto Rico								
B	Buffalo								
C	Austin								
C	Puerto Rico								
D	Toledo								
D	Puerto Rico								
E	Toledo								
E	Puerto Rico								
F	Dallas								
G	Toledo								
H	Austin								
I	North Carolina						199517		
J	North Carolina						16757		
K	Toledo								
L	Buffalo								

256

L	Lincoln	2088					
M	Lincoln	28191					
N	Lincoln	5366					
N	Toledo						
O	Lincoln	2866					
P	Park City						
Q	Park City		20007				
R	Austin						
Distribution	Dallas						
Distribution	Puerto Rico						
Manufacturing administration	Manufacturing administration						
Distribution center	Dallas 2						
Subtotal		38510				236281	
Utilization (%)		100				100	
Manufacturing (%)		100				54	
Forecasted		38510	0	0	0	292000	
Current		100000	0	100000	0	292000	56000
Additional required		−61490	0	−100000	0	145557	−56000 −8200

Table 14.2 Detailed Manufacturing Plan

Product description	Quantity/week forecast	Subassembly ratios			Forecast		
		A	B	C	A	B	C
Gadget	2500.00	1.00	0	1.25	2500.0	0	3125.0
Widget	850.00	1.00	0.50	0.75	850.0	425.0	637.5
Gizmo	1200.00	2.00	2.00	3.00	2400.0	2400.0	3600.0
				Total quantity	5750.0	2825.0	7362.5

Operation	Standard hours per subassembly			Standard hours required per subassembly		
	A	B	C	A	B	C
Prepare	0.0035	0.0055	0.0075	20.1	15.5	55.2
Assemble	0.0100	0.0250	0.0155	57.5	70.6	114.1
Clean	0.0050	0.0095	0.0750	28.8	26.8	552.2
Inspect	0.0025	0.0035	0.0450	14.4	9.9	331.3

Microcomputer Applications

Package	0.0070	0.0085	0.0056	40.3	24.0	41.2
Equipment standard hours	Factors (%)		Equipment hours	Available hours	Required hours	
	Performance	Uptime				
90.9	95.0	90.0	106.3	40.0	2.7	
242.2	95.0	75.0	340.0	40.0	8.5	
607.8	95.0	80.0	799.7	40.0	20.0	
355.6	95.0	75.0	499.1	40.0	12.5	
105.5	95.0	87.0	127.6	40.0	3.2	

Labor standard hours	Labor productivity factors	Total labor hours	Available hours	People required
1402.0	Performance = 95.0	2395.7	40.0	59.9
	Utilization = 70.0			
	Attendance = 88.0			

$$\frac{\text{Projected revenue (column 2)}}{\text{1982 Revenue per square foot (column 3)}} \times \frac{\text{Space productivity factor (column 7)}}{\text{Space utilization factor (column 6)}}$$

$$= \text{Projected square feet}$$

Beyond column 12, there is a column designated for each present and proposed production facility. In this case, there was no effort made to allocate production lines to various proposed facilities. Therefore, in some columns, the report shows the proposed facilities empty while existing facilities are projected to require much more space.

Rows 31 through 36 summarize and manipulate the total square feet for all facilities (column 12) and each separate facility. If a particular facility cannot be utilized fully (i.e., too many physical constraints, awkward shape, multilevel, etc.), a utilization factor can be entered by facility (row 32). Likewise, in a facility that is only partially dedicated to manufacturing, the manufacturing space can be adjusted by a percentage (row 33). The additional space needs, in total (column 12) and by facility, are calculated by subtracting the adjusted total space projection by the facilities' current available space.

Table 14.2 is an example of a detailed planning system done for one production line on a spread sheet program. The program generates manpower, equipment, and space requirements for a subassembly line from a top-level product forecast. Although a microcomputer-based spread sheet is limited for detailed planning, it can be effective at modeling a relatively simple product and process.

15
Productivity Planning: Utilizing the Manufacturing Planning System

15.1 INTRODUCTION

The focus of productivity planning is on improving the way things operate, whereas manufacturing planning focuses on determining the assets that are needed to operate. Although productivity planning includes more than asset effectiveness, the manufacturing planning approach can provide the foundation for productivity planning.

Beyond predicting future asset requirements, a manufacturing planning system can significantly contribute to any company's productivity improvement efforts. Since a manufacturing plan quantifies the relationship between the product and the process, a planning system can be used to evaluate an operation's current and future productivity.

Section 5.4 has been reprinted with permission from *Industrial Engineering* magazine, Vol. 15, No. 2, February 1983. Copyright © Institute of Industrial Engineers, 25 Technology Park/Atlanta, Norcross, GA 300925; (404) 449-0460.

A planning system can be used as a tool to conduct productivity audits. As an auditing tool, the planning system becomes a control system for the present rather than a planning tool for the future. Auditing the manufacturing operation will uncover specific areas requiring improvement. It will uncover areas where expectations are not being met and areas where expectations are not high enough.

Beyond monitoring productivity, a manufacturing planning system can be used as an analytical tool. With an effective planning system, any number of variables can be altered to evaluate their impact on the operation's productivity.

A manufacturing planning system can become more than a tool for auditing and analysis. It can become the centerpiece for a major corporate comprehensive productivity improvement effort. A planning system can help eliminate the "piecemeal" approach to productivity improvement programs.

Although a planning system can take many forms, the approach described in this text can provide a comprehensive yet consistent link between all facets of a productivity improvement program. It can support a productivity drive by providing a tool for auditing, analysis, and planning.

15.2 AUDITING PRODUCTIVITY

The technique for utilizing the manufacturing planning system for auditing productivity is the same as that used for long-range planning. The major difference is that instead of using a marketing forecast, the planner must input a current production schedule. An indication of an operation's productivity can be seen in the difference between what the system says the manufacturing requirements should be and the requirements that were actually used. For instance, the system may state the production schedule could be met with 100 people. In actuality it may have taken 120 people. The beauty of a planning system is not so much that it points out the discrepancy, but that it can be used to discover what caused that discrepancy. The problem could range from too many people in a particular work center, to worse than expected absenteeism.

The planning system can provide a means for comparison between calculated requirements and actual requirements. As such, it is a useful tool for keeping production management honest. It can act as a checks and balances to ensure that

Auditing Productivity

assets on hand are the assets that are needed. Motivated by the need to meet shipping schedules, it is sometimes in the production manager's best interest to have excess capacity.

A manufacturing planning system can provide a means for monitoring productivity in terms of assets rather than dollars. A planning system can provide emphasis on productivity where it belongs: on the management of the operations and not on the manipulation of the finances.

As an auditing tool, the planning system has one flaw. If the standards are loose, it is possible to appear productive even though that appearance is not deserved. With loose standards, the productivity factors are easy to control. Utilization, performance, and uptime can all be made to look good. The calculated assets and those on hand may match. However, if both the actual and calculated assets are based on loose standards, the auditing function is meaningless.

Auditing the standards must be part of any comprehensive productivity audit. Even if the standards are not changed as a result of the audit, the performance factor, or a productivity improvement factor, should be adjusted to reflect a more reasonable expectation.

Standards that are not maintained typically deteriorate. Standards can become inaccurate due to a number of factors. Over time, even small product and process improvements can render a standard obsolete. Although standards are best audited and maintained continuously, sometimes a short-term quick audit is necessary.

Auditing the standards does not necessarily mean auditing each individual standard for each individual product at each individual operation. The planner is unlikely to single-handedly update the entire standards file. An audit of the standards should be conducted at a level of detail that is applicable to the planning system.

Typically, rather than update every standard, the performance factor is used to adjust the total standard hours required for an operation. Most likely a planning system will not utilize a separate performance for each product going through each operation. Therefore, the standards that are audited should be representative of the product mix that typically passes through that operation. One approach is to audit the standards for the two or three items that make up the bulk of the volume going through that operation. The other approach is to randomly choose a few items for audit. With either approach, the point is to determine any discrepancy between the task as it is documented and the task as it is actually performed.

Another method of auditing standards is to compare time study data with predetermined times. If the standards were established through the use of a stopwatch, it may be helpful to review the same job with a predetermined time system and vice versa.

Once the standards have been audited, it is then possible to monitor the maintenance of those standards. The manufacturing planning system can be used as a standards audit control system. The standards audit control system works by using a fixed production schedule and fixed productivity factors. The only variable that is allowed to change is the industrial engineering standards file. The system is then run once a month or once a quarter to generate the manufacturing assets report. That report is then compared with the previous report or the initial report. Any disparity can be attributed only to the standards.

Hopefully, throughout the year there is a reduction in the number of standard hours required to produce the same amount of product. The rate of improvement can then be tracked in order to establish a productivity improvement factor for future planning. If, over time, more standard hours are required for the same production schedule, there may be a problem. However, the increase in standard hours may be due to an engineering change order, adherence to tighter quality standards, or perhaps a change in method where some indirect labor is now done by direct labor. Although such a change would suggest an increase in direct manufacturing costs, there may be trade-offs.

Although usually not seen in a positive light, an increase in standard hours could be a productivity improvement—an improvement in the sense that some additional production time may significantly reduce test and repair time. On a broader scale, customers who are pleased with a well-made product will tend to buy more of that product. There may be sound business reasons for taking additional time to build it right the first time.

Whether the trend in standard hours is going up or down, it is important to know the trend. Only through routine standards audits will this vital information be known in a timely fashion.

15.3 ANALYZING PRODUCTIVITY IMPROVEMENT POTENTIAL

The manufacturing planning system can be an analytical tool. It can be used to calculate and quantify the impact of a variety of productivity improvement efforts.

The planning system provides a means for predicting the consequences of hypothetical improvements in productivity. The planning system is particularly good at presenting the impact that a small improvement, or series of improvements, will have on the big picture. For instance, many manufacturers establish productivity improvement or cost reduction goals for the year. Attaining these goals is usually dependent on a number of product and process changes. Only through a planning system can the cumulative impact of these changes be easily assessed. For example, the system is capable of generating an assets plan based on no improvements. The results can then be compared to a plan that is based on a multitude of improvements. The difference will effectively illustrate the magnitude, and the cumulative effect, of a number of small improvements. Also, the small improvements can be manipulated to achieve the desired end result. Not only can the system determine the overall improvement in productivity, but it can specifically demonstrate how the overall improvement can be achieved.

As an analytical tool, the planning system can be effective for evaluating alternative operating scenarios. Manufacturing is like many problems where there is no one correct answer. It is entirely possible to get the product out the door through a variety of methods. The planning system may be especially effective at analyzing the productivity of manufacturing assets based on these alternative methods, although it must be recognized that the productivity of assets may be only one of many relevant criteria when evaluating alternatives.

The planning system should not be confused with a sophisticated simulation system. While useful for analyzing "what if" scenarios, it does no simulation based on time-phased probability or random occurrences. The planning system, as described in this text, is a relatively simple number cruncher.

15.4 PLANNING PRODUCTIVITY

The manufacturing planning approach may also be a useful tool for productivity planning. It can help establish productivity goals and monitor progress toward those goals. A documented manufacturing plan can become the foundation of a comprehensive productivity improvement program.

15.4.1 Productivity in Perspective

Before embarking on a productivity planning program, care must be taken to place "productivity" in the proper perspective. This can be done by understanding productivity in the context of the entire organization and its relationship to sound business practice. Four principles establish this perspective:

1. Improved productivity is the result of any activity that reduces costs in the long run without compromising the corporate philosophy and the quality of the product or service, or reducing gross revenue. Too often productivity is related only to reducing costs. There are other considerations as well: product quality, management philosophy, marketing strategy, social concerns, and environmental concerns. For example, it does no good to "improve productivity" by reducing product quality, threatening the work force, ignoring safety, or polluting the environment. While these actions may reduce costs in the short term, they could decimate revenues in the long term.
2. Planning productivity is an action, not a reaction. Many productivity improvement programs are implemented as a means to catch up with a changing business environment. This reaction only enhances the already prevalent short-range mentality of management, which usually insists on a short-term payback for productivity improvement projects. Equal emphasis should be placed on planning for the future and avoid having to catch up later. The lack of productivity programs is especially pronounced in companies experiencing a period of success and rapid growth. At this stage, profitability does not depend on productivity, since the company sells whatever it can make at a

profit. This is precisely the time when an ambitious effort should be devoted to productivity planning. Eventually the product growth curves will level off, whether due to product life cycle, the economy, or competition. A good example can be seen in the minicomputer industry, and more recently the microcomputer industry. The combination of these factors has led to a situation where margins are eroding and profitability now does depend very much on productivity.
3. Productivity is a universal objective, not an individual function. Productivity planning, although needing a coordinator, must be an organizational effort directed at the whole organization. An organizational effort maximizes the quantity and quality of improvement ideas, and paves the way for their implementation.
4. Productivity goals must be specific, achievable, and readily understood and supported by those charged with their implementation.

A productivity plan based on realistic and detailed objectives will greatly enhance the successful achievement of those objectives. Specified projects with implementation schedules will minimize ambiguity and enable progress to be measured. This level of detail will also enable productivity goals to be incorporated as an integral part of the company's financial, strategic, and operating plans. Then productivity, similar to budgeting, becomes a serious concern and one aspect of determining management performance.

15.4.2 An Approach to Productivity Planning

The comprehensive nature of a serious productivity improvement program dictates that all facets of a manufacturer's operation be scrutinized. Productivity planning should go beyond direct manufacturing assets and operations, which are only one component of a company's productivity picture. This perspective is needed, even though the most significant, tangible, and immediately achievable improvements will most likely be found within the manufacturing operation.

The process of developing a comprehensive productivity plan can be outlined in seven steps:

1. Establish priorities: attack areas with the most improvement potential.
2. Document the present operation.
3. Visualize the future operation assuming no improvements.
4. Document applicable technologies, techniques, and trends.
5. Develop a scenario for the future, assuming improvements are applied.
6. Set goals and objectives to realize the improvement scenario.
7. Implement the plan and monitor progress.

15.4.3 Prioritizing Improvement Potential

Obviously, productivity planning for the entire organization is a major undertaking. To facilitate the initial success and effectiveness of the effort, it should be directed at those areas where the benefits will be most apparent. Unfortunately, the areas with the highest costs are not necessarily the areas where the most gains can be made. Prioritizing areas for further productivity planning requires not only financial and strategic considerations, but also consideration of improvement potential.

Uncovering those areas with the greatest potential for improvement is a manageable task. It can be done with the systematic procedure outlined below. A format will be established for displaying where both asset and operational costs are incurred. In a matrix form, this format will also display how much is spent. Productivity potential is then estimated and a percentage entered for each item in the matrix. Finally, using current costs and productivity potential, the potential dollar savings can be calculated. These saving estimates will highlight those areas with the greatest improvement potential and point to strategy to achieve potential.

Design a Cost Matrix

One method of identifying areas with the greatest potential for productivity planning is to look at the costs that cut across organizational lines. Similar functions and expenses may respond to the same productivity improvement techniques. For example,

Planning Productivity

a variety of functions, such as marketing, finance, research and development, and production control, all utilize analysts and clerical personnel. Although each department is independent, the analyst and clerical personnel engage in similar tasks and follow similar procedures. By identifying the costs and improvement potential of these personnel as a group, rather than by function or department, an area of major potential can become a priority for productivity planning.

Table 15.1 demonstrates two ways to categorize a company's costs, in this case, a manufacturer. The vertical categories designate traditional profit centers or departments. The horizontal categories represent assets, functions, or other expense areas that cut across organizational lines. In application, this matrix would be broken down into a finer level of detail.

The input for the categories will have to come from upper management, department heads, a financial analyst, and a productivity specialist (i.e., industrial engineer, organizational development specialist, or internal consultant). The categories and their level of detail must be such that actual costs and the impact of productivity improvements are relatively easy to approximate. The idea is to get a rough estimate of where the preliminary efforts should be directed, not to establish specific goals. Examples can be seen in Table 15.1.

Establishing the categories and designing the matrix are the first steps of the productivity planning process. Enlisting the cooperation and support of the management and their staff at this early date may be crucial to the program's success.

Enter Annual Costs (Table 15.1)

The manager of each department, utilizing his or her annual budget history, should be asked to report a cost figure for each of the items designated on the horizontal axis. Depending on the detail of some of the items and their variance from budgeting categories, some assistance may be required from the finance department or a productivity specialist. Assistance from these sources also will help insure consistency throughout the organization. In this example, the highest costs can be attributed to the hourly and direct labor groups, at $3.39 and $3.35 million/year. The bulk of these figures is being generated by the hourly workers doing direct labor, at a cost of $2.75 million/year. Direct labor cost is the primary area of expenditure, followed by the cost of analysis and clerical work at $2.05 million and direct material costs of $2.0 million.

Table 15.1   Annual Costs in $10,000s

Direct costs	Direct material (200)	Space	Energy	Hourly labor
	Direct labor	–	–	275
Factory overhead	Maintenance department	5	5	40
General overhead	Research and development department	5	10	4
	Finance department	5	5	5
Sales costs	Marketing department	5	5	5
	Sales department	5	5	0
Total costs		25	30	339

Productivity Potential (Table 15.2)

A list of the categories from across the top of the matrix should be presented to each manager and his or her staff. Working with them, the planner could generate potential improvement ideas that could reduce the annual costs for each category. Sketchy ideas will suffice; therefore, the subject is well suited for a "brainstorming" session. The improvement ideas and suggestions are likely to fall into two groups: technologies and techniques. The productivity specialist may want to initiate discussion by submitting some examples for discussion such as the following:

    Technologies
      Robotics
      Microcomputers

Planning Productivity

Material handling	Analysis and clerical	Engineers	Information systems	Petty cash	Total cost
45	15	–	–	–	335
0	5	5	2	3	65
1	25	50	10	5	110
–	50	–	15	10	90
–	50	–	15	10	90
–	50	10	15	25	110
46	205	65	57	53	1,000

    Conveyors
    Software packages
    Teleconferencing

Techniques
    Management by objectives
    Quality circles
    Profit sharing
    Incentive pay
    Organization change

    The ideas generated must be evaluated to ensure that they are consistent with corporate philosophy and operating strategy. Some consideration should also be given to the practicality of the expense involved in their implementation. From this initial study, a list of those ideas that are deemed feasible in each category may be prepared.

Table 15.2 Rough-Cut Cost (%)

Direct costs	Direct material (20%)	Space	Energy	Hourly labor
	Direct labor	–	–	20
Factory overhead	Maintenance department	10	10	10
General overhead	Research and development department	0	10	0
	Finance department	0	10	50
Sales costs	Marketing department	0	10	20
	Sales department	0	10	–

A rough estimate of the cost reduction potential for each idea should then be calculated by the department manager and productivity specialist. For each category, the improvement ideas should be summed and a total theoretical cost reduction percentage calculated. Again, sophisticated analysis is not necessary at this stage: an educated guess is sufficient.

After collecting similar percentages from every department, the planner then can develop a new table. It should contain the best estimate of productivity improvement potential for each department and category. For example, Table 15.2 illustrates that a major improvement could be made in the cost of direct labor for material handling. This cost reduction could come from conveyors, robotics, or other mechanical handling systems. In another instance, the finance department estimates that the money devoted to simple number crunching by analysts and clerks could be cut significantly through the use of personal desk top microcomputers. This, combined with a quality circles program, has the potential of reducing analytical and clerical costs by 50%, while at the same time eliminating boring tasks.

Material handling	Analysis and clerical	Engineers	Information systems	Petty cash
50	20	–	–	–
–	20	0	0	0
0	50	0	10	0
–	50	–	33	10
–	50	–	33	10
–	50	0	0	20

Potential Cost Avoidance (Table 15.3)

By using the annual costs on Table 15.1, multiplied by the estimated cost reduction percentages on Table 15.2, a cost avoidance matrix can be calculated. This third table would define in dollars the estimated costs that would be avoided had the ideas for improvement been in place. Categories with the most potential for productivity improvement would be highlighted on a completed matrix (Table 15.3). This approach will remove from further study those categories and areas with no initial ideas for improvements. Those areas that show not only great potential for cost savings, but also a rough idea of how to achieve that potential will be prioritized for productivity planning.

As can be seen in Table 15.3, the greatest potential for cost reduction lies in the costs attributed to "analysis and clerical" work. Hourly labor, although a primary expense category in Table 15.1, has a productivity improvement potential that is second to both the analysis and clerical category, and the direct labor group as a whole. Therefore, the category of analysis

Table 15.3 Potential Cost Avoidance in $10,000s

Direct costs	Direct material (40.0)	Space	Energy	Hourly labor
	Direct labor	–	–	55.0
Factory overhead	Maintenance department	0.5	0.5	4.0
General overhead	Research and development department	–	1.0	–
	Finance department	–	0.5	2.5
Sales costs	Marketing department	–	0.5	1.0
	Sales department	–	0.5	–
Total cost		0.5	3.0	62.5

and clerical work should receive top priority for productivity planning.

15.4.4 Understanding the Present

Once a category or function has been identified as having potential for productivity improvements, the task of detailed planning can begin. The manufacturing planning techniques described in this text provide the foundation for productivity planning.

The best place to start is with the development of a comprehensive and accurate picture of the current operation. A portrayal of the present operation serves three purposes: (1) to ensure that productivity goals are based on a realistic understanding of the current situation; (2) to develop a basis of

Material handling	Analysis and clerical	Engineers	Information systems	Petty cash	Total cost
22.5	3.0	–	–	–	80.5
–	1.0	–	–	–	6.0
–	12.5	–	1.0	–	14.5
–	25.0	–	5.0	1	34.0
–	25.0	–	5.0	1.0	32.5
–	25.0	–	–	5.0	30.5
22.5	91.5	–	11.0	7.0	198.0

comparison once productivity plans and improvements are implemented; and (3) to ensure that goals are specific and achievable.

As with manufacturing planning, the process flow chart accompanied by a data sheet is one of the best tools for documenting current operations. In a manufacturing environment, the chart serves to clarify understanding of the operation and ensure that nothing is overlooked. Each step of the process is detailed including inspections, transport steps, various delays, and storages (see Fig. 15.1). Values can then be attached to each step denoting a wide range of theoretical and actual indices of productivity. Through these charts and figures, an accurate picture of present operating conditions can be established.

Part of understanding the present is to recognize its future consequence if nothing is improved. An effective method for understanding the impact of continuing "as is," and highlighting

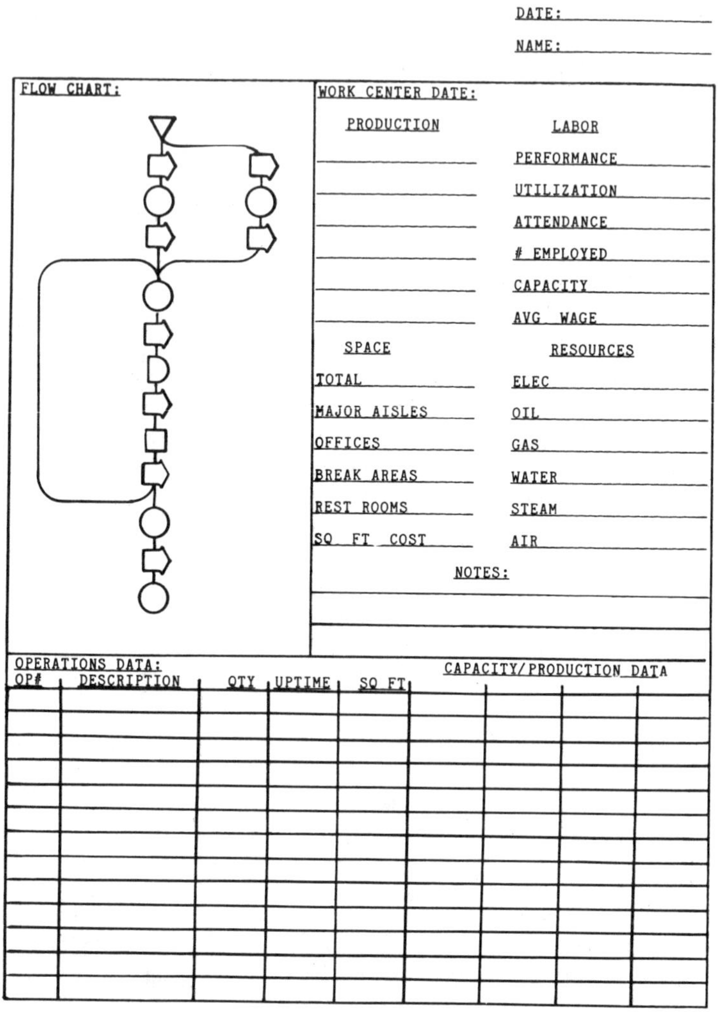

Fig. 15.1 Process flow chart and data sheet.

potential for improvement, is to envision the future operation based on today's operating conditions.

Utilizing the picture of the current operation "as is," and a marketing forecast, the data sheets can be revised to reflect the resources necessary to achieve that forecast. This process could be accomplished easily through the use of a computer-aided manufacturing planning system. The new picture that emerges illustrates how the operation will look in the future, assuming no improvement in productivity. It presents a worst-case scenario on which to establish and measure the potential impact of productivity improvements. It also helps to ensure that a long-range perspective is maintained.

Although the manufacturing environment is the thrust of this text, the same technique could be applied to other productivity studies. A flow chart and other data could just as easily be created to illustrate energy usage throughout a building, information flow, customer service routines, or even the tasks of analysts and clerical workers. A projection of resource requirements could then be done based on no change in how they function. Later a "gap analysis," comparing this worst-case scenario, will enable productivity benefits to be quantified and specific goals established.

This picture of the future will be used to assess the impact of a productivity plan. Rather than emphasizing a short-term payback, emphasis is placed on the far-sighted cost avoidance benefits of medium- to long-term productivity planning.

15.4.5 Techniques, Technologies, and Trends

Research is the key to planning for the future. The suggested improvement ideas must be investigated. These ideas and a number of other factors and future trends will impact productivity. These factors and trends must be identified and their potential impact evaluated. This evaluation is manageable since most productivity improvements being factored into a 5-year productivity plan will be based on incorporating techniques, technologies, and trends that are currently documented.

Technology

The primary contributor to great strides in productivity will be technology. In a recent article in *Modern Office Procedures*,

John J. Connell, President of Office Technologies, states: "The potential of the marketplace and the forces of competition will result in the introduction of a bewildering array of machines with constantly expanding capabilities. One can attempt to manage these technologies in an intelligent way or one can let them pour in haphazardly."

To manage intelligently, a list should be made of all currently and soon to be available technologies applicable to the area being studied. For each item on the list, detailed documentation of applicability and productivity potential can be collected from trade magazines, journal articles, and recent books. The documentation should include facts and figures on capabilities, tolerances, capacity, costs, advantages, disadvantages, etc. Realistic productivity goals can then be established based on the success of previous installations.

Techniques

The same procedure can also be followed for investigating various management techniques and assessing their improvement potential. Determining the feasibility of quality circles, employee participation, or incentive systems requires research into similar industries and applications.

Trends

It is not only necessary to identify the potential for various technologies and techniques, but also their compatibility with socioeconomic trends that may impact the enterprise, its products, and the ways it conducts its business. Successful planning for the future requires an understanding of trends that may change the nature of business and the structure of the working environment.

Significant demographics and other trends are important and necessary considerations in any productivity plan. For instance, changes in workers attitude, career planning, and education level may require a new approach to job design. As career expectations rise, the work force becomes increasingly discontent with mundane, repetitive work. Therefore, understanding this trend would dictate that the traditional industrial engineering practice of segmenting jobs may no longer be the most beneficial approach. Rather, techniques that address the human factor would be more productive and compatible with this trend.

Planning Productivity

Through this analysis of technologies, techniques, and trends, a usable data base has been gathered. This information can then be used to manage the future.

15.4.6 Managing the Future

The future can be managed and controlled with an effective productivity plan that is aggressively incorporated into the business' operating and strategic plans. This approval requires productivity goals that are specific, achievable, and readily understood and supported by those charged with their implementation. Productivity goals, based on technologies and techniques compatible with future trends and the operating philosophy, should become an integral part of the company's operation.

Defining the Future Potential

Potential can be illustrated by developing a detailed scenario that presents an identifiable, concrete end result of productivity planning. This new scenario, illustrating the improved operation in 5 years, also can be used to quantify the impact of productivity planning by comparing this scenario with the picture of the operation assuming no improvement. Measuring productivity against the future, rather than the present, also serves to insure a commitment to long-range thinking.

Two sources of data, developed earlier, will be used to construct this new scenario: (1) the flow charts and other data that illustrate the operation 5 years ahead with no productivity improvement, and (2) the documentation on technologies, techniques, and trends. Combining the two, a scenario will emerge illustrating the potential of the operation in 5 years.

This new scenario must be developed in detail. Each step and aspect of the operation needs to be analyzed and factored to reflect the successful application of technologies and techniques.

Realizing the Potential

As was stated earlier, practical productivity planning should specify achievable goals and document the means of achieving them. Once a detailed scenario for the future is prepared, an implementation schedule or project plan must be developed. This schedule will establish benchmarks and target dates for the completion of tasks throughout the 5-year planning period. Benchmarks will serve to promote a sense of accomplishment, while also measuring progress.

Annual objectives then can be set based on these benchmarks. Assuming that the objectives are realistic and supported, commitment to their achievement should be total. The company's financial, strategic, and operating plans should also incorporate these objectives. Productivity would then attain the status it deserves with management, becoming a consideration in day-to-day decision making and assuring that every effort is made toward its success. The more productivity is incorporated into business planning, the more seriously it will be taken. Therefore, it is more likely that the scenario for the future will be achieved.

Like any business planning, productivity planning cannot be done once every 5 years and be effective. It will require constant modification and reevaluation. In addition, adjustments will have to be made to adapt to changing business conditions.

15.4.7 Summary

A corporate commitment to productivity planning, combined with a practical approach, can insure future profitability. Productivity planning improves decision making, optimizes future operational effectiveness, and uncovers current inefficiencies.

This planning process, implemented with the same emphasis as budgeting, is an effective tool for serious attack on the productivity problem. As this process is implemented and goals established, it can then be incorporated into the company's strategic, financial, and operating plans. Successfully integrated into the company's operation, productivity planning simply becomes good management.

Index

Aisles, 110, 115
Attendance, 127-128

Boulton, W., et al., 10

Capacity
  planning, 2
  requirements, 165
Computer
  application, 83-86, 235, 242-247
  feasibility, 235-238
  files, 102
  hardware, 241-242
  languages, 240-241
  mainframe, 242-244

[Computer]
  micro, 244-246
Connell, J., 278
Corporate philosophy, 20-22, 36-37
Cost
  avoidance, 3, 273-274
  budgeting, 227-274
  of current operations, 269-270
  of planning, 8, 43-44, 51-52
Cost reduction, 139
CPM (critical path method), 230

Data
  collection guidelines, 66

[Data]
  constant change, 61–62
  contradictions, 62–63
  documentation, 63–64
  importance of, 59–60
  organization of, 64–65
  process, 91–92
  product, 67–68
  timeliness vs. accuracy, 60–61
  variable vs. fixed, 65
  warehouse, 194
DeLorean Motor Company, 15
Digital Equipment Corp., 15
Documentation, 63–64, 223–226

Engineering
  industrial, 50-51, 54
  manufacturing, 137
  studies, 136–138
Equipment
  "footprint," 110–113
  hours available, 175
  lead time, 16
  planning data, 104–108
  process time, 104–106
  projected, 174–177, 183–187
  reported systems, 135

Facility data, 109
Fitzpatrick and Puttock, 28
Forecast
  add-on options, 71–73, 162
  components/subassemblies, 157–161
  factoring, 161–162
  marketing, 68–70
  smoothing, 154–157

[Forecast]
  spares, 70–71
  time frame, 153–157

Gantt chart, 229–230

Hales and Muther, 10
Hayes and Wheelwright, 25
Hill, 68
Holden, P., et al., 2
Honeywell, Inc., 15

Industrial engineering, 50–51, 54
Information systems, 134–135
Integrated operations planning, 20, 29
Integration, vertical vs. horizontal, 23–25
Inventory
  finished goods, 213
  level of, 198–208
  raw materials, 211–213
  seasonal adjustment, 203–204
  work-in-process, 114

Labor
  administrative, 191–192
  current, 103–104
  planning data, 99
  projected, 169–174, 181–183, 190–192
  reporting systems, 134
  standards, 99–103
    computer filed, 102

Index

[Labor]
    estimated, 103
    standards audit, 263–264
    time available, 153–154
    unapplied, 126–127
    warehouse, 213–215
Lead time offset, 163–164
Long-range planning
    history of, 9–10

Management
    corporate, 36–38
    finance, 6
    manufacturing, 5, 40–42
    marketing, 5–6, 38–40
    personnel, 5
    time, 42–45
Manufacturing
    overseas, 14
    process steps, 91–92
    relationship to marketing, 39
    strategy, 22–23, 29
Manufacturing assets planning
    definition, 1
    need for, 10–17
Manufacturing engineering, 137
Marketing
    forecast, 68–70
    relationship to manufacturing, 39
    organization, 38–39
Material handling, 106, 110
Muther, R., 147

Niland, P., 2

Office space, 118–121
Ouchi, W., 21

Pallet space, 195–198
Performance
    equipment, 125–126
    labor, 124–126
PERT, 229–230
Planning
    activities, 56–57
    auditability, 46
    budget, 227–229
    capacity, 2
    coordinator, 53–54
    cost of, 8, 43–44, 51–52
    evaluation of, 217–219
    facilities, 2, 231–234
    flexibility of, 45, 48
    history of, 9–10
    implementation, 227–231
    objectives of, 35
    organization for, 54–56
    process of, 141
    production, 2
    productivity, 3, 261–262, 266
    revisions, 219–223
    rough-cut, 146–150
    schedule, 229–231
    tactical, 142–143, 217
    time frame, 151–153
    users and uses, 4–7, 48–51
    warehouse, 193, 210
Planning matrix, 167
Planning systems, 32
    flexibility of, 48
Plant layout, 98
    block plans, 233
    equipment "footprint," 110–113

Polaroid, 22
Presentations, 223-227
Process data sheets, 96-98
Process flow chart, 92-96
Process lead time, 163-164
Process yield, 130-132, 168-169
Product
　complexity, 13-14
　configuration, 76
　configuration tracking system, 83-86
　design, 87-89
　dynamic configuration, 78-83
　life cycle, 13, 25
　new, 86-87
　options, 71-73
　secondary, 70-73
　static configuration, 76-78
Production
　reports, 136
　units of, 73-75
Production management
　goals, 138-139
Production reports, 136
Productivity
　audits, 262-264
　definition of, 266-267
　improvement, 7, 98-99, 131-133, 137, 234, 265, 270-272
　planning, 261-262, 266-268
Productivity factors
　definition of, 123-124
　historical trends, 140
　improvements in, 138-139
　purpose, 123

Rework, 130-131

Scrap rate, 130-131
Shapiro, 39
Site selection, 231-233
Space
　equipment "footprint," 110-113
　indirect, 114-118, 178-180, 190-192
　pallet "footprint," 196-198
　projected, 175-178, 187-192
　utilization factor, 117-118, 190
Spare parts, 70-71, 161-162
Strategy
　corporate, 20-22
　manufacturing, 22-23
Systems
　reporting, 134-135

Tactics
　determination of, 145, 219-222
Technology, 12, 30-31, 277-278
Time
　availability for production, 153-154
　management of, 42-45
Toffler, A., 14
Tegoe and Zimmerman, 21
Trends, 29, 278
　historical, 140

Uptime, equipment, 128-130
Utilization
　labor, 126-127
　space, 117-118

Wang Laboratories, Inc., 15

Index

Warehouse
  data, 194
  finished goods space, 213–215
  planning, 193, 210
  raw materials space, 211–213
  support areas, 210

Work center, 178–179
Work force, changes, 16

Yield, process, 130–132
  168–169